WITHDRAWN

RTC Limerick

D1582407

Molecular Thermodynamics

MOLECULAR THERMODYNAMICS

An Introduction to Statistical
Mechanics for Chemists

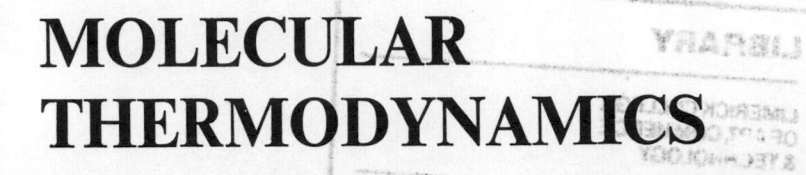

John H. Knox
Professor of Physical Chemistry,
University of Edinburgh

Revised Edition

John Wiley & Sons
Chichester · New York · Brisbane · Toronto

LIBRARY

LIMERICK COLLEGE
OF ART, COMMERCE
& TECHNOLOGY

class no: 536·7 KNO
acc.no: 7743

Copyright © 1971, 1978, by John Wiley & Sons, Ltd.

All rights reserved.

No part of this book may be reproduced by any means, nor transmitted, nor translated into a machine language without the written permission of the publisher.

Library of Congress Cataloging in Publication Data:

Knox, John H.
  Molecular thermodynamics.

  Bibliography: p.
  Includes index.
  1. Statistical thermodynamics. I. Title.
QC311.5.K54 1978   536'.7   70–147399

ISBN 0 471 99621 1

Printed in Great Britain by Unwin Brothers Limited
The Gresham Press, Old Woking, Surrey, England
A member of the Staples Printing Group

iv

PREFACE TO REVISED EDITION

In preparing this edition of *Molecular Thermodynamics* I have taken the opportunity to make a number of minor corrections and to clarify the text in places where it appeared obscure. I have also made substantial alterations to the introductory Chapter 1 and to the treatment of hindered internal rotation in Chapter 9. Otherwise the scheme of the book remains unchanged, namely to present a fairly detailed general derivation of the main concepts of molecular thermodynamics in Chapters 1–5 and to apply them to practical examples in the second part, Chapters 6–12.

The numerical examples at the end of each Chapter, which have not been changed, will it is hoped enable the student to test his understanding of the material in each Chapter.

J. H. KNOX
Edinburgh
October, 1977

PREFACE

Chemistry deals ultimately with molecules and their interactions, but its observations are almost invariably made on matter in bulk. A lengthy extrapolation and much imagination has been required to interpret these observations in atomic or molecular terms. Experience has shown that matter observed in bulk exhibits two distinct kinds of property. Properties of the first kind reflect the co-operative actions of large numbers of molecules, and are observed when all contributing molecules undergo the identical process. Molecular phenomena can then be seen on a greatly intensified and amplified scale but relatively undistorted. Spectroscopic properties are the prime examples of this class, and can nowadays be interpreted directly in terms of the properties of individual molecules. Properties of the second type arise when different molecules provide different contributions to the observed property. The values of such properties reflect averages taken over all the molecules under observation. Typical of such properties are thermodynamic properties such as internal energy, entropy, heat capacity, pressure, dielectric constant. There is no immediate way of linking the molecular properties which can be derived from the first type of bulk property, and the thermodynamic properties of the second type. A connection must nevertheless exist for ultimately the thermodynamic properties of matter must depend upon the properties of the constituent molecules.

The subject which provides the link is the subject of this book, and is variously called statistical mechanics, statistical thermodynamics, molecular statistics, statistical physics, or molecular thermodynamics. Its foundations were laid in the nineteenth century by Maxwell, Boltzmann and Gibbs, but important advances were made in the present century with the application of the ideas of quantum mechanics. This greatly simplified the foundations of the subject and extended its usefulness. Molecular thermodynamics is now an indispensable part of physical chemistry.

The present book is intended as an introduction which will be suitable for second and third year undergraduates of chemistry who have taken a course in classical thermodynamics, and who may be assumed to be familar with differential and integral calculus. I have not assumed any knowledge of quantum mechanics apart from a familiarity with the general idea of quantization.

Since the concepts of quantum and classical mechanics are most important for the development of molecular thermodynamics, I have devoted Chapter 2 to an elementary account of them. The exposition has strictly limited aims, and is not meant to be regarded as a proper introduction to quantum or classical mechanics *per se*. Readers who are already familiar with this material may feel that they can omit this section. Chapter 3 gives a brief summary of classical thermodynamics whose aim is to expose the logical foundations of the subject, and to outline the main conclusions of chemical significance.

Chapter 4 is the core of the first part of the book, and is inevitably the most difficult one. The foundations of molecular thermodynamics can be laid in several ways, none of which are particularly simple. I have avoided the conventional method of development which is based upon a consideration of an assembly of independent systems, and employs the mathematical method known as "Lagrange's method of undetermined multipliers". To my mind this method of approach obscures the fact that only a single assumption, in addition to those of mechanics and thermodynamics, is required to develop the entire subject. It also has the more serious drawback that it is applicable only to assemblies of independent systems (ideal crystals and gases) for which well defined molecular quantum states exist. Although we finally derive relationships only for these ideal assemblies, the treatment given in Chapter 4 is general for all types of assembly, and so may be used as a starting point for the development of theories of non-ideal assemblies such as real gases and crystals. It is therefore hoped that this introduction can be used to give a direct entry to more advanced texts, and that the student will not require to start afresh before he can treat more advanced topics.

Chapter 4 first develops the idea of the assembly partition function, Q, and then shows how the thermodynamic properties of an assembly may be obtained in terms of Q. Only in Chapter 5 is the specialization to assemblies of independent systems made, and formulae developed for thermodynamic properties in terms of the molecular partition function. In Chapters 6 and 7 molecular partition functions for translation, rotation and vibration are evaluated, the quantum formulation in Chapter 6, and the classical formulation in Chapter 7. Chapters 8 and 9 deal with the contributions to the thermodynamic properties of ideal crystals and ideal gases from the different modes of motion.

In Chapter 10, the Maxwell–Boltzmann distribution law is developed and applied. Chapters 11 and 12 conclude the book with consideration of chemical equilibrium and reaction rate theory.

Since the theory of molecular thermodynamics is justified entirely by the validity of the results it predicts, I have tried to include adequate illustrative examples which compare values of thermodynamic parameters found by experiment, with those calculated from molecular parameters using the equations of molecular thermodynamics. These examples should also clarify the methods used to calculate thermodynamic parameters.

Following current recommendations, I have used S.I. (Système International) units throughout the book. The definitions of these units are fully described in the booklet produced by the Royal Society of London *Symbols, Signs and Abbreviations* 1969. The definition of the "mole" as the gram mole instead of the kilogram mole results in some confusion which I have tried to circumvent by using N_A and L_A for Avogadro's number expressed respectively in systems per mole and systems per kilomole.

In writing this book I have become much indebted to many previous authors, but particularly to Fowler and Guggenheim (*Statistical Thermodynamics*, Cambridge University Press, 1952), Andrews (*Equilibrium Statistical Mechanics*, Wiley, 1963), Rushbrooke (*Statistical Mechanics*, Oxford University Press, 1941) and Eyring, Henderson, Stover and Eyring (*Statistical Mechanics*, Wiley, 1964). Most of the worthwhile ideas which appear in the present work arise from them.

I am also indebted to my students who, over the past few years, have indicated, in the clearest possible ways, where arguments and deductions given in lectures have been unsound, unclear, or incorrect.

J. H. KNOX
March, 1971

RECOMMENDED VALUES OF PHYSICAL CONSTANTS

The following values are taken from *Symbols, Signs and Abbreviations 1969* published by the Royal Society, London, and have been approved by the International Unions of Pure and Applied Physics and Chemistry.

Quantity	*Symbol*	*Values with estimated uncertainty*
Speed of light	c	$= (2 \cdot 997925 \pm 0 \cdot 000003) \times 10^8$ m s^{-1}
Permittivity of a vacuum	ε_0	$= (8 \cdot 854185 \pm 0 \cdot 000018)$ $\times 10^{-12}$ J^{-1} C^2 m^{-1}
Mass of hydrogen atom	m_H	$= (1 \cdot 67343 \pm 0 \cdot 00008) \times 10^{-27}$ kg
Mass of electron	m_e	$= (9 \cdot 1091 \pm 0 \cdot 0004) \times 10^{-31}$ kg
Charge on proton	e	$= (1 \cdot 60210 \pm 0 \cdot 00007) \times 10^{-19}$ C
Boltzmann constant	k	$= (1 \cdot 38054 \pm 0 \cdot 00018) \times 10^{-23}$ J K^{-1}
Planck constant	h	$= (6 \cdot 6256 \pm 0 \cdot 0005) \times 10^{-34}$ J s
	h/k	$= (4 \cdot 7993 \pm 0 \cdot 0006) \times 10^{-11}$ s K
Avogadro constant*	N_A	$= (6 \cdot 02252 \pm 0 \cdot 00028) \times 10^{23}$ mol^{-1}
	L_A	$= (6 \cdot 02252 \pm 0 \cdot 00028) \times 10^{26}$ kmol^{-1}
Ice-point temperature	T_{ice}	$= (273 \cdot 1500 \pm 0 \cdot 0001)$ K
Faraday constant	$F = N_A e$	$= (9 \cdot 64870 \pm 0 \cdot 00016) \times 10^4$ C mol^{-1}
Gas constant	$R = N_A k$	$= (8 \cdot 3143 \pm 0 \cdot 0012)$ J mol^{-1} K^{-1}

Subsidiary standards

Standard atmosphere	1 atm	$= 1 \cdot 01325 \times 10^5$ N m^{-2} exactly
Electron volt	1 eV	$= 1 \cdot 6021 \times 10^{-19}$ J approximately
Atomic mass unit	1 amu $=$	
	$1/N_A$	$= 1 \cdot 66041 \times 10^{-27}$ kg approximately
Inch	1 in	$= 2 \cdot 54 \times 10^{-2}$ m exactly
Calorie (thermochem.)	1 cal	$= 4 \cdot 184$ J exactly

Numbers

π	$= 3 \cdot 141593$
e	$= 2 \cdot 7182818$
ln 10	$= 2 \cdot 302585$

*The mole is defined as the quantity of material containing the same number of systems as there are atoms in exactly $0 \cdot 012$ kg of the isotope carbon-12.

CONTENTS

xiii

PART II PARTITION FUNCTIONS AND THEIR APPLICATIONS

Part I

FUNDAMENTALS

CHAPTER 1

INTRODUCTION

Molecular thermodynamics forms the link between quantum mechanics and classical thermodynamics, and it is an integral part of modern physical chemistry. Its key position in chemistry is seen by considering the scope and limitations of the two disciplines which form its boundaries.

Classical thermodynamics does not concern itself with molecules, and the molecular idea appears incidentally and peripherally because chemists have found that the laws of chemical combination take particularly simple forms when the quantities of chemical substances are expressed in moles. Indeed the simplicity introduced into chemistry by the use of the mole concept is one of the compelling arguments for the atomic or molecular nature of matter. Thermodynamics is based upon four laws of experience which between them define the ideas of temperature, internal energy and entropy, and which specify the essential features of the absolute scales of temperature and entropy. The strength of thermodynamics lies in the generality of its laws, and in the large number of valuable relations which arise from their mathematical formulation. Typical of such relationships is that between the equilibrium constant, K_p, and the enthalpy change, $\Delta H°$, for a chemical reaction:

$$\left(\frac{\partial \ln K_p}{\partial T}\right)_p = \frac{\Delta H°}{RT^2} \tag{1.0.1}$$

The inevitable weakness of thermodynamics is its inability to comment on the connection between molecular and bulk parameters.

Quantum mechanics, on the other hand, deals almost exclusively with matter at the atomic or molecular level. It is based upon a small number of formal mathematical rules, complementing those of Newtonian physics, which enable the results of classical mechanics to be applied to atomic sized particles. The rules incorporate the idea of the wave–particle duality of atomic matter. They enable the permissible stationary states of atomic–molecular systems (electrons, protons, neutrons, quanta of

3

radiation, atoms, molecules, etc.) to be described by wave functions which are functions of spatial coordinates but not of time: they represent stationary waves in space. According to quantum mechanics, the wave functions give the entire information about the permissible states of systems to which they refer. Amongst the most important information is the energy of each stationary state, ϵ, and the probability of finding a system in a given volume around any coordinate. This is proportional to the square of the wave function at the particular coordinate.

The forms of the wave functions for any system, say an electron, are determined entirely by the constraints which would be imposed upon the motion of the system if it obeyed the laws of classical mechanics. For instance the wave functions for the electron in the hydrogen atom are determined by the rules of quantum mechanics and the necessity that the force between the electron and the proton obeys Coulomb's law, the two particles being regarded as point masses. The wave function for a particle free to move within a rectangular box is determined by the dimensions of the box. The complete set of wave functions which are allowed to a system under a given set of constraints is called an "ensemble" of wave functions, and this set embodies all possible information about the behaviour of the system under these constraints.

Quantum mechanics, however, gives no indication as to which wave function from an ensemble will represent the state of a system at any instant. Nor does it give any information about the general state of a group of similar systems, say 10^{20} hydrogen molecules at 300 K, or 10^{20} atoms of gold in a crystal of gold.

We observe that neither classical nor quantum mechanics is able to tell us how to obtain the thermodynamic properties of matter in bulk from the properties of the molecules themselves. Yet it is obvious that if matter is indeed molecular, and the evidence for this is overwhelming, we must be able to derive the one from the other. This conviction existed long before the advent of quantum mechanics, and even before the advent of classical thermodynamics, being exemplified in the kinetic theory of Bernouilli as early as 1738. This theory, the forerunner of the modern kinetic theory of gases, gave the first convincing explanation of the pressure–volume relationship for a gas. Two important assumptions were made by Bernouilli: first, that the molecules of a gas were in constant random motion, and second, that the pressure of a gas resulted from the averaging of the effects of myriads of molecular impacts on the walls of the container. Later, correlation of the predictions of the kinetic theory with experimental observations in the form of the ideal gas law, the law of effusion, and other relationships, led to the conclusion that the temperature of a gas was proportional to the average kinetic energy of the molecules, the proportionality factor being the same for all molecules irrespective of their masses. Temperature measured on the "ideal gas scale" was called the "absolute temperature". The unit of temperature on the ideal gas scale is called the Kelvin and is defined so that the absolute temperature of water at its triple point is exactly 273·15 K.

For our purpose it is important to observe that, even in the earliest kinetic–molecular theories, bulk properties like pressure are obtained from individual molecular contributions by averaging over all molecules; other properties like temperature, which have no meaning at the molecular level, can be identified with average properties only by making *reasonable* assumptions. Functions must therefore be developed by the statistical treatment which we can by hindsight identify with thermodynamic parameters.

The successful development of methods for correlating molecular and thermodynamic parameters was accomplished in the latter part of the nineteenth century by Maxwell, Boltzmann, Gibbs and others. Their work was a major intellectual triumph. Classical statistical mechanics accepted the then universally held view that molecules behaved in the same way as massive bodies, and obeyed the laws of Newtonian mechanics. Around the turn of the century it began to become apparent that this view was incorrect. In thermodynamics the third law was discovered which suggested strongly that the entropies of perfect-crystals at the absolute zero of temperature were zero. Classical statistical mechanics gave no indication of any such zero of entropy. In molecular theory the work of Planck, Einstein, Bohr, de Broglie, Heisenberg, Schrödinger and others showed just how far the physics of atomic matter deviated from that derived by Newton, and led to the development of quantum mechanics. Its application in statistical mechanics led to great simplification of the latter's conceptual basis, and at the same time solved the problem of the vanishing entropy of perfect crystals at absolute zero. It also cleared up many other anomalies, until then unexplained, such as the failure of diamond to show a molar heat capacity in accord with Dulong and Petit's law.

Although we shall develop the ideas of quantum and classical molecular thermodynamics in parallel in this book, the quantum formulation is much the simpler, and we suggest that those studying the subject for the first time should pass over the classical theory until they have read through the arguments of the quantum-based theory.

1.1 DEFINITIONS

Before outlining briefly the general line of attack on the problem of how we can link thermodynamic and molecular parameters we must define a number of terms which are widely used in molecular thermodynamics.

To avoid confusion the word SYSTEM is used exclusively for particles to which the laws of classical or quantum mechanics are applied. "System" is therefore used as a general term for an electron, proton, neutron, atom, ion, molecule, etc. It is never used in the sense of a "thermodynamic system" which has macroscopic dimensions. This is always called an ASSEMBLY. An assembly is therefore composed of systems. One mole of gaseous hydrogen is an assembly containing about 6×10^{23} systems, the systems being hydrogen molecules. A crystal of sodium chloride contains equal numbers of two types of system, Na^+ ions and Cl^- ions.

There are two important types of assembly for which, and only for which, simple statistical mechanical treatments can be given. An ASSEMBLY OF LOCALIZED SYSTEMS is one in which the systems are associated with fixed positions within the assembly, one system to each position. A crystalline substance is such an assembly. In the simplest type of assembly of localized systems, each system is independent of its neighbours, and so has its own *private set* of quantum states whose energies are unaffected by the condition of neighbouring systems. This is an idealization and may be called an "assembly of independent localized systems". Such an assembly may also be called an "assembly of private systems". If the systems are private in this sense it actually makes no difference whether they are localized or not, but this is a subtle point which we shall not develop further here. For the time being we regard both of these assemblies as the same and take the ideal crystal as the prime example.

The second type of assembly is called an ASSEMBLY OF NON-LOCALIZED SYSTEMS. It is one in which *many* systems are associated with the same region in space, and in which no system is associated with any particular part of the region over any period of time. A gas falls into this category. In quantum mechanical terms the ideal gas contrasts with the ideal crystal in that the quantum states for the translational motion of the molecules of an ideal gas are held in common by all the systems of the assembly. There are no longer private sets of quantum states for each system of the assembly. A liquid falls into neither category, for over short periods of time the systems of the assembly are localized, but over longer periods they are non-localized. For this reason the statistical mechanical treatment of liquids is extremely complex, and far beyond the scope of this book.

The word ENSEMBLE has already been introduced. In molecular thermodynamics it is used to mean an imaginary collection of states, usually quantum states, which have certain features in common. For instance, the ensemble representing the quantum states of the hydrogen atom would contain two members for the $1s$ states (one for each possible electron spin state), two members representing the $2s$ states, six members representing the $2p$ states and so on. Any real H-atom would then be in a state represented by one and only one member of the ensemble.

An ensemble may represent all the accessible states of a single molecule or system, or it may represent all accessible states of an assembly, for quantum mechanics makes no distinction in principle between a single molecule and a group of 10^{26} molecules. If an assembly is completely isolated from the outside world it has a completely constant energy which cannot change with time. The ensemble appropriate to this type of assembly then contains members only for those assembly quantum states which have a precise energy, E. It does not contain any members representing states of different energy. At any instant the real assembly will be in a quantum state represented by one and only one member of the ensemble. Gibbs called this type of ensemble, appropriate to an isolated assembly, a "micro-canonical ensemble".

When an assembly is placed in a situation where it can exchange energy with its surroundings, for example in a thermostat, it may in principle have any energy, and all quantum states whatever their energy are accessible. The appropriate ensemble then contains one member for every quantum state of the assembly whatever its energy. It contains an infinite number of members, and is called a "canonical ensemble".

Generalization may be carried one stage further by allowing the assembly to exchange not only energy but also systems with the outside world. In thermodynamic terminology such an assembly would be called an "open system". The appropriate ensemble now contains one member for every possible quantum state whatever the energy, and whatever the number of systems, and is called a "grand canonical ensemble".

Molecular thermodynamics may be developed using any of the three types of ensemble. The commonest methods use the micro-canonical ensemble, but at some stage an awkward and usually unconvincing transition has to be made to the canonical ensemble, for physical scientists are rarely interested in the properties of assemblies with which they can make no contact. In the present treatment we avoid this transition by developing the subject using the canonical ensemble. Thus, we assume from the start that assemblies of interest are those which are in thermal contact with their surroundings.

1.2 OUTLINE OF A METHOD OF PROCEDURE

In describing the kinetic theory of gases (p. 5) it was noted that while some bulk properties, such as pressure, could be obtained by direct averaging, other properties such as temperature could only be derived by making reasonable connecting assumptions. It is not difficult to appreciate that if we require four laws of thermodynamics to define the concepts of temperature, internal energy, entropy, and the entropy/temperature scale, we shall probably require four corresponding connecting assumptions to pass from the properties of atomic molecular systems to those of matter in bulk.

Since the bulk properties are obtained by some kind of averaging, it is also necessary to introduce an entirely new assumption foreign to both atomic physics and thermodynamics in order to define the way in which the averaging is to be carried out. This is the assumption that quantum states of a system (or assembly) which have the same energy are equally probable. A more rigorous statement of this assumption is given in Section 4.1.

To illustrate where the fundamental assumption is brought in and how the connecting assumptions are made we consider in a non-rigorous way a specific and limited example. Suppose we have an assembly of N independent localized systems (say an ideal crystal of argon) where each system has its own private set of quantum states and energy levels. According to quantum mechanics we can specify the overall state of this assembly by a statement such as:

The system at the first site is in a quantum state of energy ϵ_5,
the system at the second site is in a quantum state of energy ϵ_2,
the system at the third site is in a quantum state of energy ϵ_{10}, and so
on for all N sites in the crystal.

This defines what is called a complexion or an overall quantum state of
the assembly, but from the thermodynamic point of view it is unneces-
sarily detailed for it makes no difference to the observable bulk properties which site has which energy; indeed all that is important is the total
energy of the assembly, not how it is parcelled out. The fundamental
assumption of molecular thermodynamics states that all such complexions of the same energy are equally likely (Section 4.1). From this
assumption it follows (Section 4.2) that the probability of a system
being in a quantum state of energy ϵ_1 depends only upon the energy ϵ_1,
and upon a parameter β which can be shown to be an inverse measure
of the temperature of the assembly. This is the first connecting link and
corresponds to the zeroth law of thermodynamics. The relationship is

$$\left\{ \begin{matrix} \text{Probability, } p_i, \text{ that a system} \\ \text{is in a quantum state, } i, \text{ of} \\ \text{energy } \epsilon_i \end{matrix} \right\} = (1/q) \exp\left[-\beta\epsilon_i\right] \qquad (1.2.1)$$

In order that p_i as defined by 1.2.1 is a genuine probability, the sum
of all p_i must be unity, that is

$$p_1 + p_2 + p_3 + \ldots = \sum_{\substack{\text{all} \\ \text{States}}} p_i = 1 \qquad (1.2.2)$$

Thus

$$q = \sum_{\substack{\text{all} \\ \text{States}}} \exp\left[-\beta\epsilon_i\right] \qquad (1.2.3)$$

The total energy of an assembly of N such systems is then

$$E = N(p_1\epsilon_1 + p_2\epsilon_2 + \ldots)$$
$$= (N/q) \sum_{\substack{\text{all} \\ \text{states}}} \epsilon_i \exp\left[-\beta\epsilon_i\right] \qquad (1.2.4)$$

This E is now identified with the thermodynamic internal energy, U,
by the second connecting assumption which corresponds to the first
law of classical thermodynamics. By considering a specific example,
say an ideal crystal, the molecular thermodynamic expression for E can
be compared with the classical thermodynamic expression for U. This
comparison results in the identification of β

$$\beta = 1/kT \tag{1.2.5}$$

where k is the Boltzmann constant. Using 1.2.4 and 1.2.5 it is not difficult to show that

$$E = NkT^2(\text{d} \ln q/\text{d}T) \tag{1.2.6}$$

Since q is obtained from purely molecular parameters we have now shown how one of the major thermodynamic functions can be calculated from molecular data.

The correlation of β with $1/kT$ and the final correlation which we now discuss correspond to the closely related second and third laws of classical thermodynamics.

To make the final connecting link we have to consider the key property, entropy. How is this to be visualized in molecular thermodynamics? In most thermodynamic treatments the entropy is associated with the degree of disorder of an assembly or with the lack of information we have about it. It is reasonable to suppose that if we could quantify this "lack of information" we might be able to give a value to the entropy.

We might proceed as follows: in our assembly, which contains systems at N sites, an energy ϵ_1 might be associated with N_1 sites, an energy ϵ_2 with N_2 sites and so on, but we have no idea and, what is more, little interest in how the different energies might be distributed throughout the sites of the assembly. There will, of course, be a large number of possible arrangements of the energies amongst sites which will have the same total energy, E, and will be strictly equivalent from the thermodynamic viewpoint. The fundamental assumption of molecular thermodynamics asserts that all such arrangements, or complexions as they are usually called, are equally probable. The number of complexions is found by application of the theory of permutations, and equals the number of distinguishable arrangements of N objects of which N_1 are of one sort, N_2 of another sort, and so on. This number is given by equation (1.2.7)

$$W = N!/(N_1! \, N_2! \ldots) \tag{1.2.7}$$

where $N! = N(N-1)(N-2)\ldots 3.2.1$. The larger W is, the less information we have about which sites have which energies, and it is reasonable to regard W as an appropriate measure of the lack of information we have about the assembly since the actual state could be any one of the W possible states. How then might W be related to the entropy of the assembly?

We know from thermodynamics that the total entropy of two independent assemblies A and B is the sum of their individual entropies. But since every complexion of A may be taken with every complexion of B, the total number of complexions for the combined assemblies is the product of the numbers for the individual assemblies. We therefore have the two relations

$$S_{(A + B)} = S_A + S_B \qquad (1.2.8)$$

$$W_{(A + B)} = W_A \times W_B \qquad (1.2.9)$$

The only functional relationship which satisfies (1.2.8) and (1.2.9) is

$$S = k \ln W \qquad (1.2.10)$$

where k is a constant. Surprisingly, k in equation (1.2.10) has precisely the same value as the k in equation (1.2.5). It is the universal molar gas constant, divided by Avogadro's number; $k = R/N_A$. Equation (1.2.10) represents the fourth and final connecting assumption necessary to marry classical to molecular thermodynamics.

The entropy S can now be expressed in terms of q. First we state Stirling's approximation for $\ln N!$ which is valid when N is large (see page 81):

$$\ln N! = N \ln N - N \qquad (1.2.11)$$

Using (1.2.11) for the factorials in equation (1.2.7) and substituting $\ln W$ into (1.2.10) gives after some rearrangement (and remembering that $\sum N_i = N$)

$$S = -Nk \sum (N_i/N) \ln (N_i/N) \qquad (1.2.12)$$

The ratio (N_i/N) is simply the probability that a system has an energy ϵ_i and is equal to $\exp[-\epsilon_i/kT]/q$. Insertion of this into (1.2.12) followed by some simple algebra gives finally for S

$$S = E/T + Nk \ln q \qquad (1.2.13)$$

The free energy is obtained from the thermodynamic relation $F = E - TS$ and so is

$$F = -Nk\,T \ln q \qquad (1.2.14)$$

Thus all the key thermodynamic properties can be obtained in terms of the molecular partition function and the temperature for an assembly of independent systems. In principle, then, they may be obtained directly from molecular parameters. This is the essential achievement of molecular thermodynamics, but on the way it provides much insight and information into molecular behaviour.

The treatment given above is far from rigorous but it brings out the distinction between the two types of assumption which have to be made in linking molecular and bulk properties. The fundamental assumption which relates to the probabilities of complexions having the same energy is of a quite different character from the connecting assumptions which are necessary to link parameters derivable from the statistical treatment with the classical thermodynamic parameters temperature, internal energy and entropy.

The rigorous treatment given in Chapters 4 and 5 depends upon similar assumptions, but deals with a more general type of assembly and does not require the systems to be independent. It is then found that relationship (1.2.10) is no longer the best starting point for making the link corresponding to the second law, but arises as a consequence of more general equations.

PROBLEMS

1.1   Contrast, and comment on the connection between, the ensemble of quantum states accessible to H_2 molecules in a volume V, and the quantum states which are occupied by the molecules of an assembly occupying the same volume.

1.2   A particular assembly of localized systems contains three identical systems which can exist in any four quantum states whose energies are ϵ, 2ϵ, 3ϵ and 4ϵ. (a) How many assembly quantum states are there altogether? (b) What kind of ensemble do they constitute? (c) Which states constitute an ensemble for an assembly with an energy 6ϵ? (d) What kind of ensemble do they constitute?

1.3   Prove equation (1.2.6) by writing out q as a series and differentiating term by term with respect to T.

1.4   Prove equation (1.2.7) by a logical argument. Start by proving the case where all objects are different.

CHAPTER 2

QUANTUM AND CLASSICAL MECHANICS

2.1 QUANTUM MECHANICS

Experiments with fundamental particles, atomic particles and molecular systems show that such systems possess the properties of both waves and corpuscles. This duality is expressed by the Heisenberg uncertainty principle which states that it is impossible simultaneously to have precise information about the position and momentum of any particle. The coupled uncertainties are related by the equation (2.1.1)

$$\Delta p \times \Delta q \approx h \qquad (2.1.1)$$

where Δp is the uncertainty in the momentum p, Δq the uncertainty in the position q, and h is the Planck constant:

$$h = 6 \cdot 6256 \times 10^{-34} \text{ J s} \qquad (2.1.2)$$

p and q may in fact be any pair of "conjugate" quantities: if p is the normal momentum of a system, mass $\times$ velocity, then q is the Cartesian coordinate of its position; if p is the angular momentum of the system, $p_\theta = I\omega$ (I = moment of inertia, ω = angular velocity), then q has to be the angular position of the system, q_θ, measured in radians.

The wave–particle duality is also expressed in the exact relation of de Broglie, which states that a particle of linear momentum $p = mv$ is associated with a wave of wavelength λ given by equation (2.1.3).

$$p = mv = h/\lambda \qquad (2.1.3)$$

This relation allows the total energy ϵ of the particle to be calculated when it moves in a field of uniform potential energy, for then

$$\text{Total energy} = \text{kinetic energy} = \frac{mv^2}{2} = \frac{p^2}{2m} = \frac{h^2}{2m\lambda^2} \qquad (2.1.4)$$

Two examples of systems to which the de Broglie relation can be directly

13

applied are a particle described classically as moving in a one-dimensional box of length L (i.e. along a line of length L), and a particle described classically as moving in a circle of radius r.

In classical terms a particle constrained to move within a one-dimensional box of length L will move back and forth along a line of length L with a definite velocity v. It may in principle have any velocity v. In quantum mechanical terms it must be described by a stationary wave of a certain wavelength. By the uncertainty principle if one is certain about its wavelength and hence its momentum, one can have no certainty about its position, Thus it is best to imagine the particle as spread out over the whole line, and represented simply and entirely by the wave which fills the length L. For the wave to be stationary it must have nodes at the ends of the line in the same way as a vibrating violin string has nodes at the finger and at the bridge. Some of the various possible standing waves are shown in Figure 2.1. The condition

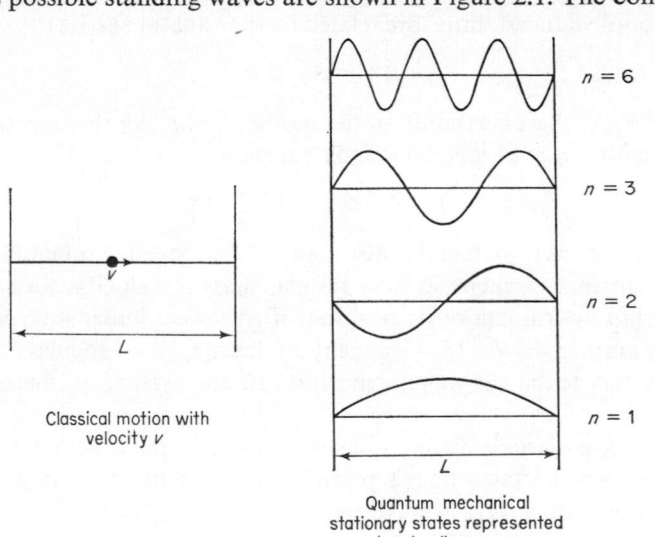

Classical motion with
velocity v

$n = 6$

$n = 3$

$n = 2$

$n = 1$

Quantum mechanical
stationary states represented
by standing waves

Figure 2.1 Classical and quantum mechanical representations of one-dimensional translation: $n =$ quantum number.

for a wave to be stationary is simply

Length of box = integral number of half wavelengths

$$L = n \times (\lambda/2), \quad n = 1, 2 \ldots \tag{2.1.5}$$

Eliminating λ by applying the de Broglie equation (2.1.3) gives

$$p = hn/2L \qquad (2.1.6)$$

from which the energies of the stationary states are obtained as

$$\epsilon = h^2n^2/8mL^2 \qquad (2.1.7)$$

For a particle described classically as moving in a three-dimensional rectangular box of sides a, b and c, the total energy is the sum of three terms, one for each independent component of the translational motion:

$$\epsilon = \frac{h^2}{8m}\left\{\frac{j^2}{a^2} + \frac{l^2}{b^2} + \frac{n^2}{c^2}\right\} \qquad (2.1.8)$$

where j, l and n may take any positive integral values apart from zero. Other energies are not allowed since they would not correspond to standing or stationary waves.

For a particle described classically as moving in a circle of radius r, the quantum condition for it to be represented by a stationary wave is that the total circumference of the circle must contain an integral number of wavelengths as shown in Figure 2.2. If this condition is not

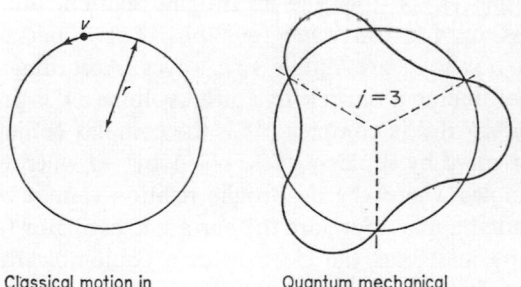

Classical motion in
a circle of radius r

Quantum mechanical
stationary state represented
by a standing wave

Figure 2.2 Classical and quantum mechanical representations of one-dimensional rotation: j = quantum number.

satisfied destructive interference occurs. The stationary wave condition is then

$$2\pi r = j\lambda, \quad j = 0, 1, 2, \ldots \qquad (2.1.9)$$

Following equations (2.1.3) and (2.1.4) the energy is given by

$$\epsilon = \frac{h^2 j^2}{8\pi^2 m r^2} = \frac{h^2 j^2}{8\pi^2 I} \tag{2.1.10}$$

where $I = mr^2 =$ moment of inertia of particle about the axis of rotation.

When the potential energy of the particle regarded classically depends upon the coordinates of the particle, the de Broglie relation cannot be applied directly, and the more general Schrödinger equation must be used. This equation incorporates the de Broglie relation, and has the form

$$-\frac{h^2}{8\pi^2 m}\left\{\frac{\partial^2\psi}{\partial x^2} + \frac{\partial^2\psi}{\partial y^2} + \frac{\partial^2\psi}{\partial z^2}\right\} + U(x, y, z)\,\psi = \epsilon\psi \tag{2.1.11}$$

$U(x, y, z)$ is the potential energy which the particle would have if treated classically, and is a function of the particle's position only, The solutions to this equation are expressions for the total energy, ϵ, and for the wave functions, ψ. The wave functions depend upon the coordinates x, y and z. The energies, called eigenvalues of the equation, on the other hand, are functions only of molecular parameters; they do not involve the coordinates x, y and z. When applied to situations for which $U =$ constant, the Schrödinger equation gives the same values for the energies, ϵ, as does the de Broglie relation, but, in addition, gives expressions for the wave functions. For a particle in a one-dimensional box they are simple sine waves. According to Born, the probability of finding a particle in a given volume $\mathrm{d}V$ is proportional to $\psi^2\,\mathrm{d}V$, or $\psi\psi^*\mathrm{d}V$ if ψ is complex (ψ^* is the complex conjugate of ψ and is obtained from ψ by replacing $i = \sqrt{-1}$ by $-i$ wherever it occurs).

Two examples where the de Broglie relation cannot be applied, or has only limited application, are the harmonic oscillator (for example a diatomic molecule) and the electron in a coulombic force field (for example the electron in the H-atom or He$^+$ ion). The forms of the potential function for these and for the particle in a one-dimensional box are shown in Figure 2.3.

Solutions to the Schrödinger equation give the energies of the allowed quantum states of the harmonic oscillator as

$$\epsilon_n = (n + \tfrac{1}{2})h\nu \tag{2.1.12}$$

where ν is the fundamental oscillation frequency, and n any integer including zero.

The energy levels allowed for a single electron attracted to a nucleus of charge ze where e = electronic charge in coulomb are

$$\epsilon = -\frac{mz^2e^4}{8\epsilon_0^2h^2n^2} \tag{2.1.13}$$

where ϵ_0 is the permittivity of a vacuum; $\epsilon_0 = 8\cdot8542 \times 10^{-12}\,\text{J}^{-1}\text{C}^2$ m^{-1} and n is any positive integer.

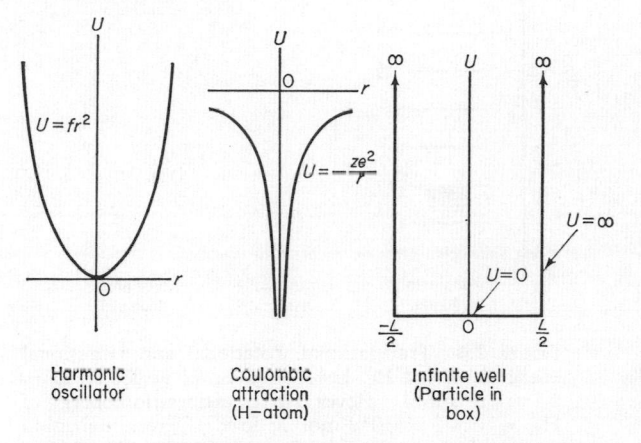

Figure 2.3 Potential functions for various types of system.

The forms of the wave functions and their energies depend strongly upon the shape of the potential function and the mass of the particle. Speaking qualitatively the smaller the "box" within which a particle is constrained, and the lighter the particle the farther apart are the energies of the allowed quantum states. Figure 2.4 shows a logarithmic plot of the energy levels for translation, rotation and vibration in the H_2 molecule assumed to be contained in a cube of side 10^{-2} m. For translational motion where the "box" size is of the order of 10^{-2} m the energy levels are closely spaced and around 300 K there is a very large number of levels with energies below kT (Note $k = R/N_A$, where R = gas constant, N_A = Avogadro's number and $k = 1\cdot3805 \times 10^{-23}$ J K^{-1}). For rotational motion the "box" size is equivalent to the circumference of the circle of gyration, that is about 10^{-9} m. The spacing of the energy levels is correspondingly greater, and there are normally only a few levels whose energies are below kT. For vibrational motion the "box" size may be taken as the amplitude of the vibration. This

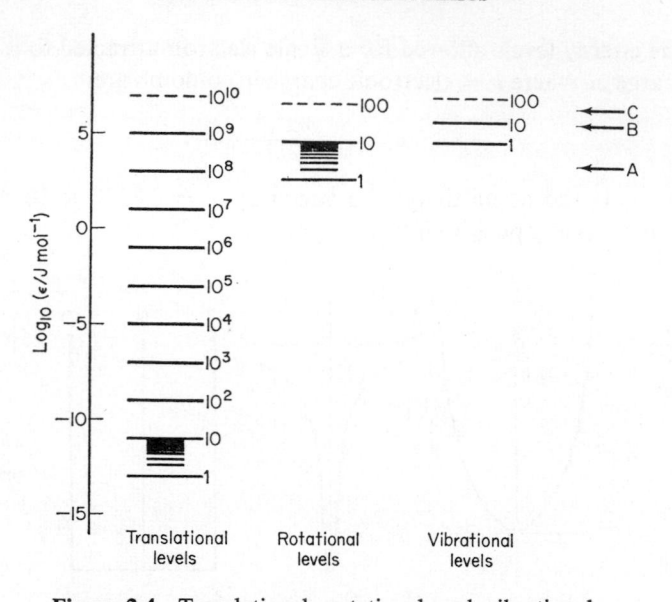

Figure 2.4 Translational, rotational and vibrational energy levels of H_2. Level A gives RT at 300 K, $\epsilon_A = 2\cdot5 \times 10^3$ J mol^{-1}; level B gives dissociation energy of H_2, $\epsilon_B = 4\cdot3 \times 10^5$ J mol^{-1}; level C gives ionization energy of H-atom $\epsilon_C = 3\cdot2 \times 10^6$ J mol^{-1}. The energy levels denoted by ------ are imaginary since H_2 dissociates below this energy.

varies with the energy of the oscillator but is of the order of 10^{-11} or 10^{-12} m. Consequently the vibrational energy levels are farther apart than the rotational levels, and the spacing is generally large compared to kT at 300 K.

An important consequence of the dependence of the energies of quantum states upon "box" size is that the energies for single systems or for assemblies of systems remain constant only if the volume within which the system or assembly is constrained remains constant. Since any mathematical formulations are greatly simplified if the energies of quantum states can be taken as invariant, processes like differentiation and integration will always be performed at constant volume. The main consequence of this is that the thermodynamic parameters which arise from molecular thermodynamic derivations are $U, F (= U - TS)$ and C_v associated in classical thermodynamics with processes carried out at constant volume. The more common parameters $H (= U + PV)$

$G (=H - TS)$, and C_p associated with processes carried out at constant pressure must be obtained from U, $\cdot F$ and C_v using standard thermodynamic relationships, but inevitably their direct calculation by the methods of molecular thermodynamics requires the knowledge of how the energies of the allowed quantum states depend upon volume. For gases this is readily obtained, but for solids it is not.

When discussing the implications of quantum mechanics for molecular thermodynamics, a clear distinction must be made between a quantum state and an energy level. The distinction may be clarified by reference to the states of the H-atom. The energy levels of the H-atom are given by the expression (2.1.13). For $n = 1$ there is a single solution to the Schrödinger equation, namely

$$\psi_{1s} = A_1 \exp [\alpha r] \tag{2.1.14}$$

where $\alpha = \pi m e^2/(\epsilon_0 h^2)$, and r is the distance of the electron from the nucleus. A_1 is a constant chosen so that

$$\int \int \int \psi^2 \, dx \, dy \, dz = 1 \tag{2.1.15}$$

This ensures that $\psi^2 \, dx \, dy \, dz$ is the true probability of finding the electron within the volume $dx \, dy \, dz$.

For $n = 2$ there are two *types* of solution to the Schrödinger equation and four solutions in all, one of the first type and three of the second type. All four solutions are associated with the same value of ϵ. The single solution of the first type is spherically symmetrical, like the wave function for the $1s$ state, but it includes a factor which is first order in r:

$$\psi_{2s} = A_2(2 - \alpha r) \exp [-\alpha r/2] \tag{2.1.16}$$

The three solutions of the second type are of lower symmetry, being cylindrically symmetrical about one of the cartesian axes rather than spherically symmetrical. They have the forms:

$$\left. \begin{aligned} \psi_{2p_x} &= A_2 . \alpha . x \exp [-\alpha r/2] \\ \psi_{2p_y} &= A_2 . \alpha . y \exp [-\alpha r/2] \\ \psi_{2p_z} &= A_2 . \alpha . z \exp [-\alpha r/2] \end{aligned} \right\} \tag{2.1.17}$$

In general there are n different *types* of solution for any value of n and n^2 solutions in all. All these solutions correspond to the same

2

energy ϵ_n as given by equation (2.1.13). An energy level such as ϵ_n which corresponds to several distinct quantum states is said to be *degenerate*. The degeneracy of an energy level is the number of wave functions which have the particular energy. Generally the degeneracy increases with n for any form of constraint, and the increase is the more rapid the more degrees of freedom are involved in the type of motion being considered. For one-dimensional motion the degeneracy is the same for all energy levels with the possible exception of the lowest.

In the chapters which follow it is essential to be clear about the distinction between a quantum state and an energy level. The former is never degenerate; the latter is very often degenerate.

At this point we have to admit that a further complication exists in the definition of a quantum state which affects to some extent what has been said about degeneracy. Detailed analysis of spectra and more profound wave-mechanical treatment reveals the phenomenon known as spin. All fundamental particles have a spin angular momentum component of $\frac{1}{2}(h/2\pi)$, and any fundamental particle can thus be assigned a spin quantum number of $\pm\frac{1}{2}$ and an appropriate wave function. There are thus two quantum states corresponding to each solution for the electron in the H-atom. The degeneracy of any energy level should therefore allow for the degeneracy due to the various possible spin quantum states. The degeneracies of the various electronic energy levels of the H-atom are therefore $2n^2$ not n^2 as previously suggested.

The total degeneracy of any energy level ought strictly to allow also for the degeneracy due to nuclear spin. In the majority of cases of interest in chemistry, nuclear spin factors cancel and can be ignored, but they are specifically responsible for the anomalies in the specific heat of hydrogen.

To deal with systems containing many fundamental particles (for example atoms with several electrons, molecules with several atoms or assemblies of systems, the Schrödinger equation must be elaborated and must include additional differential terms for the three cartesian coordinates of each particle in the system, or assembly.

For two identical particles it takes the form:

$$-\frac{h^2}{8\pi^2 m}\left\{\frac{\partial^2\psi}{\partial x_1{}^2} + \frac{\partial^2\psi}{\partial y_1{}^2} + \frac{\partial^2\psi}{\partial z_1{}^2} + \frac{\partial^2\psi}{\partial x_2{}^2} + \frac{\partial^2\psi}{\partial y_2{}^2} + \frac{\partial^2\psi}{\partial z_2{}^2}\right\}$$

$$+ \, U(x_1, y_1, z_1, x_2, y_2, z_2)\,\psi = \epsilon\psi \quad (2.1.18)$$

U and the solutions ψ are now functions of six spatial coordinates, x_1, y_1, z_1, for the first particle and x_2, y_2, z_2 for the second particle. The problem of finding solutions becomes increasingly difficult as the number of particles increases, and the solutions themselves become increasingly complex unless simplifying assumptions can be made. Nevertheless it is important to realize that however many particles are in a system and however many systems are in an assembly, solutions exist in principle. It is therefore quite permissible to talk in terms of the allowed quantum states of an assembly which may have macroscopic dimensions.

As the number of particles or systems rises, the number of solutions to the wave equation rises. For complex molecules and for practical assemblies, the number of solutions becomes astronomical. Under these conditions the quantum states and energy levels become so numerous and closely spaced that it becomes convenient to speak in terms of the quantum state density instead of in terms of the degeneracy of individual energy levels. The quantum state density, denoted by $N(\epsilon)$ or $N(E)$ depending upon whether we are treating a single complex system, or an assembly, is defined as the number of quantum states per unit energy range at the energy ϵ or E. There are thus $N(\epsilon)\,d\epsilon$ or $N(E)\,dE$ quantum states with energies in the range ϵ to $\epsilon + d\epsilon$ or E to $E + dE$.

What has been said above about the quantum states of an assembly is independent of whether the systems of the assembly interact or not. If they interact strongly, and so cannot be considered as independent, the potential function U and the assembly wave functions must be functions of the coordinates of all the systems of the assembly, and the problem of obtaining solutions to the wave equation is insoluble. However, if the systems of the assembly are essentially independent, the potential function may be split into a sum of potential functions, one for each system of the assembly. The Schrödinger equation then separates into a sum of independent parts, one for each system. The parts can be solved independently to give system (or molecular) wave functions and energies. Under this restriction both the quantum mechanical and statistical mechanical problem can be solved. The two important types of real assembly which approximate to this ideal are the monatomic crystalline solid and the dilute (ideal) gas. They are, of course, what have been referred to above as assemblies of independent localized, and independent non-localized systems.

When the systems of the assembly are independent and localized,

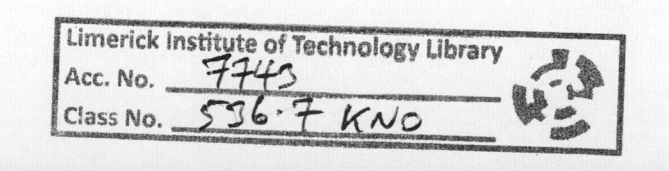

Limerick Institute of Technology Library
Acc. No. _____7743_____
Class No. _____536·7 KNO_____

which is roughly true for a crystal, a particular quantum state of the assembly may be specified by assigning specific molecular quantum states to the molecules at the specific sites in the assembly. It is important to note that although the systems of the assembly are absolutely indistinguishable (assuming they are all of the same chemical species), the sites occupied by the systems are distinct and can, in principle, be recognized by their coordinates. The wave functions for such an assembly are then found to be products of factors, one factor for each site in the assembly. Suppose for example that we have a rudimentary assembly consisting of four identical systems which occupy sites a, b, c and d. One particular state or complexion of the assembly might be described by the statement

"The system at site a is in quantum state 1

the system at site b is in quantum state 5

the system at site c is in quantum state 2

the system at site d is in quantum state 9"

The wave function which represented this particular state of the assembly would then be

$$\psi_{1529} = \psi_1(a) . \psi_5(b) . \psi_2(c) . \psi_9(d) \qquad (2.1.19)$$

The energy of this state of the assembly would be

$$E = \epsilon_1 + \epsilon_5 + \epsilon_2 + \epsilon_9 \qquad (2.1.20)$$

Another assignment of the molecular quantum states to the four sites with the same total energy would have a wave function

$$\psi_{1925} = \psi_1(a) . \psi_9(b) . \psi_2(c) . \psi_5(d) \qquad (2.1.21)$$

This is a quite distinct quantum state from the first one. It is not difficult to see that the number of quantum states or complexions which have an energy E given by (2.1.20) when all four system quantum states are different will be $4.3.2.1 = 4! = 24$. The energy level E is therefore 24-fold degenerate. If identical wave functions had been assigned to two sites the number of distinct assembly wave functions would be reduced by 2!, and the degeneracy would be $4!/2! = 12$.

As the number of sites in the assembly increases, the degeneracy of any energy level of the assembly also increases. In an assembly containing N localized systems there would be $N!$ states of a given total energy if all systems were in different quantum states. If there were N_1 systems in quantum state 1, N_2 in quantum state 2 and so on, the degeneracy of a particular energy level of the assembly would be

$$W = N!/(N_1! \, N_2! \, N_3! \ldots) \qquad (2.1.22)$$

where

$$N = N_1 + N_2 + N_3 + \ldots = \sum N_i \qquad (2.1.23)$$

$$E = N_1\epsilon_1 + N_2\epsilon_2 + N_3\epsilon_3 + \ldots = \sum N_i\epsilon_i \qquad (2.1.24)$$

W is sometimes called by the misleading name "thermodynamic probability". Since W is always much greater than unity it is in no sense a probability, but more correctly represents the number of distinguishable quantum states or complexions of the assembly for a given set of distribution numbers N_1, N_2, etc.

When several identical systems occupy the same region of space, that is when the systems are non-localized as they are in a gas, not only are the systems indistinguishable as they are in any assembly, but so are the "sites". Indeed there is only one site for all the systems of the assembly, namely the region of space within which they are all commonly contained. The indistinguishability of the "sites" is evident in the quantum mechanical treatment of systems in a box for the wave function, which represents any system, occupies the whole volume of the box. According to quantum mechanics no meaning is attached to the idea of a particle being at a particular point in the box. It then follows that no meaning can be attached to statements such as "There is a stationary state of the assembly in which

system α is in quantum state 1

system β is in quantum state 5

system γ is in quantum state 2

and so on"

All that can be said is that "There is a stationary state of the assembly in which molecular quantum states 1, 2, 5 and so on are occupied." Figure 2.5 illustrates diagrammatically the distinction between the possible quantum states of an assembly of localized and non-localized systems.

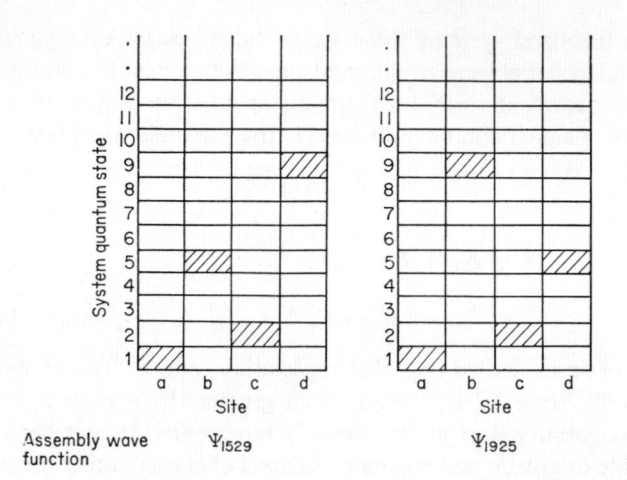

Distinguishable states of an assembly of four localized systems

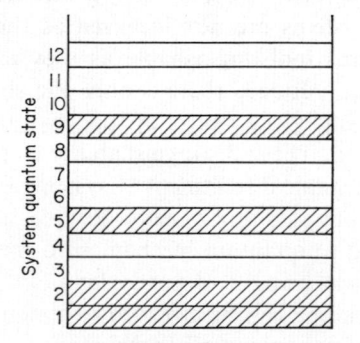

Assembly wave function equation 2.1.25

State of an assembly of four non-localized systems

Figure 2.5   States of assemblies of four systems in which molecular quantum states 1, 2, 5 and 9 are occupied.

The indistinguishability of systems of the same type is basic to quantum mechanics and has to be built into any correct version of the wave function for an assembly of non-localized systems. There are only two satisfactory ways of doing this. Suppose first of all that the systems are indeed distinguishable, say by labels; we can then write down a large number of different wave functions for any given total energy and distribution of occupied quantum states. If all systems were in different quantum states there would be $N!$ such assembly wave functions. It is then necessary to mix these $N!$ wave functions in such a way that no preference is given to any one. In mathematical terms this is achieved by writing the assembly wave function as a determinant with N rows and columns. Thus if quantum states 1, 2, 5 and 9 were occupied in a four system assembly, the assembly wave function would take the form of equation (2.1.25.)

$$\Psi_{1259} = \begin{vmatrix} \psi_1(\alpha) & \psi_2(\alpha) & \psi_5(\alpha) & \psi_9(\alpha) \\ \psi_1(\beta) & \psi_2(\beta) & \psi_5(\beta) & \psi_9(\beta) \\ \psi_1(\gamma) & \psi_2(\gamma) & \psi_5(\gamma) & \psi_9(\gamma) \\ \psi_1(\delta) & \psi_2(\delta) & \psi_5(\delta) & \psi_9(\delta) \end{vmatrix} \tag{2.1.25}$$

where $\psi_1(\alpha)$ implies that the wave function ψ_1 is associated (in imagination only) with particle α and so on. The determinant may be evaluated either with alternate $+$ and $-$ signs as is usual, or with all signs the same. Both methods preserve the indistinguishability of the systems. The rules governing which method is to be used are well understood and are as follows:

(1) The determinant is evaluated in the normal way with alternate $+$ and $-$ signs if the wave functions for the assembly are to be antisymmetric in the systems, that is if the sign of Ψ has to change when the labels on two systems are interchanged. This is the case when the typical system of the assembly contains an ODD number of fundamental particles (that is the number of electrons plus neutrons plus protons). Examples of such systems are ^{3}He, D, HD, electrons, $D^{35}Cl$, CH_3D.

(2) The determinant is evaluated with all signs the same (say positive) if the wave functions for the assembly are to be symmetrical in the systems, that is if the sign of Ψ is to remain unchanged when the labels on two systems are interchanged. This is the case when the typical system of the assembly contains an EVEN number of fundamental particles. Examples of such systems are ^{4}He, H, H_2, $H^{35}Cl$, CH_4. Photons also come into the category of particles for which the assembly wave function must be symmetric.

These rules arise because it is found that all wave functions must be antisymmetric with respect to exchange of identical fundamental particles. If the typical system contains say two fundamental particles then exchange of two systems make the assembly wave function change sign twice. In the end it does not change sign at all. If a system contains three fundamental particles then exchange of two systems causes three changes of sign in the wave function for the assembly. The final result is therefore a change of sign.

When Ψ is antisymmetric it is not possible for two systems in the assembly to occupy precisely the same quantum state, for then two columns of the determinant would be identical and Ψ would vanish. For such systems the Pauli exclusion principle applies with regard to systems. On the other hand when Ψ is symmetric no such restriction applies and multiple occupancy of quantum states becomes possible. Generally this distinction is unimportant in chemistry, but there are instances where it is vital. The unusual properties of ^{4}He as compared to ^{3}He at low temperatures results directly from this difference. The unexpected absence of any electronic heat capacity in metals arises because the electrons are forced to occupy high energy quantum states by the exclusion principle.

2.2 CLASSICAL MECHANICS

According to quantum mechanics the state of any system is completely described by the wave function for the system. Thus while the energy of a system may be sharp, its position is indeterminate. The most we can state about the position of a particle in quantum mechanical terms is the probability of finding it in a volume dV, this being proportional to the square of the wave function in that region.

In classical mechanics there is no restriction on the precision with which conjugate variables such as position and momentum can be specified. Indeed the proper classical specification of the state of any system requires exact information about *both* its position and its momentum at any instant. The classical picture of a system thus contains *more* information than the quantum mechanical picture. In attempting to correlate quantum and classical descriptions of matter the clear-cut conclusions of classical physics must be blurred. A degree of uncertainty must be introduced.

For the purpose of molecular thermodynamics, it is convenient to describe the classical motions of systems in terms of their momenta and spatial coordinates, rather than in terms of their velocities and positions. Furthermore when several particles or systems are considered we can best describe their combined motions by the motion of a single "representative point" in "phase space". Phase space is a multidimensional space in which distances along different axes give the positional and momentum coordinates of the systems of the assembly. We illustrate the idea by some simple examples.

(*a*) A particle in a one-dimensional box. The motion of a single particle back and forth along a line of length L is pictured in Figure 2.6(*a*). The instantaneous state of the particle is specified by stating

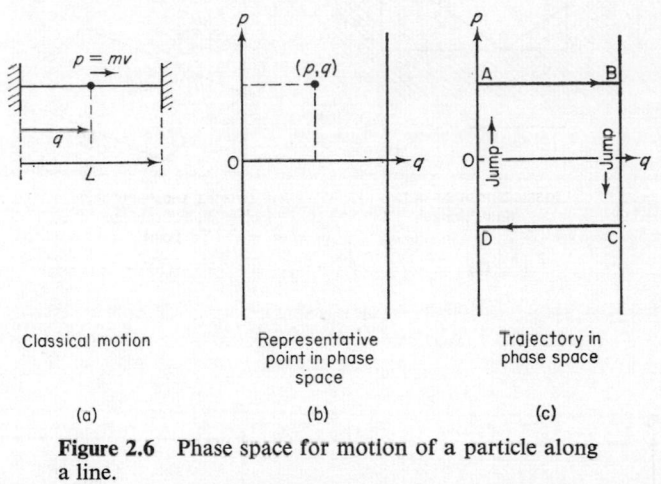

Classical motion   Representative point in phase space   Trajectory in phase space

(a)   (b)   (c)

Figure 2.6 Phase space for motion of a particle along a line.

its position, q, and its momentum, p. If two dimensions are used instead of one, we can measure q along the x-axis and p along the y-axis as in Figure 2.6(*b*). Any point in the (p, q) plane then represents a particular position and momentum of the particle, that is the instantaneous state of the particle. As the particle moves along the line in Figure 2.6(*a*) its momentum remains constant until it collides with the wall at $q = L$, whereupon its momentum changes abruptly from $+q$ to $-q$. The particle then moves back along the line until $q = 0$ when the momentum again abruptly changes sign. The representative point of the system thus executes a cycle ABCDA in Figure 2.6(*c*). The motion of the point along AB and CD is regular but the motions from B to C and from D to A are jumps. The (p, q) plane in Figure 2.6(*b*) and (*c*) is called the phase plane or the phase space for the system, and the cycle ABCDA is called the trajectory of the representative point in phase space.

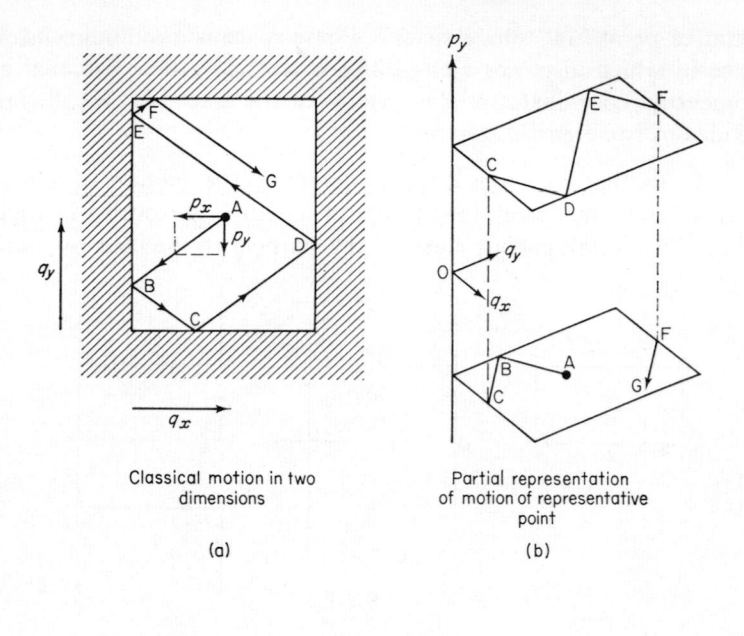

Classical motion in two
dimensions

(a)

Partial representation
of motion of representative
point

(b)

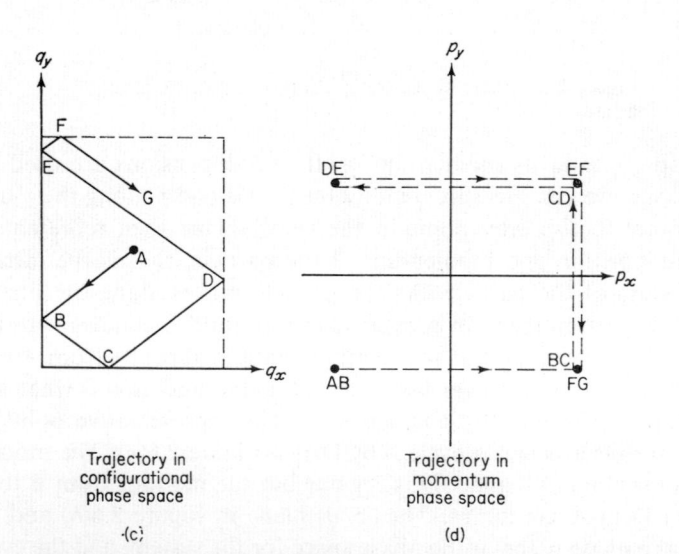

Trajectory in
configurational
phase space

(c)

Trajectory in
momentum
phase space

(d)

Figure 2.7 Phase space for motion of a particle in a plane.

To represent a particle moving throughout an area, that is in a two-dimensional box, we require a phase space of four dimensions, two for the position of the particle, and two for the components of the momentum of the particle in the two directions. Figure 2.7(a) shows the configurational trajectory of the particle and the components of the momenta in the x and y directions. We could represent say the y-component of the momentum by a third dimension as shown in Figure 2.7(b), and so give a partial representation of the motion of the representative point. However, a more satisfactory procedure is to represent the position and the momenta on separate diagrams. The motion of the representative point in *four* dimensions is now replaced by the *concerted* motion of two representative points, one in configurational phase space (two-dimensional) and one in momentum phase space (two-dimensional). As the point in configurational phase space moves regularly throughout the area shown in Figure 2.7(c), the point in momentum phase space jumps between the four points shown in Figure 2.7(d). The jumps occur when the particle hits a wall, and one of the momentum components changes sign.

To represent the motion of a single particle in a three-dimensional box requires a six-dimensional phase space which can be divided into a three-dimensional configurational phase space and a three-dimensional momentum phase space. The movement of the representative point in the former will be smooth, while that in the latter will be a series of jumps.

(b) The phase space for a one-dimensional rotator, that is a particle rotating in a circle, requires two dimensions, one, q_θ, to represent the configuration of the system, that is the orientation of the particle relative to a fixed direction in the plane, and the other to represent its angular momentum, $p_\theta = I\omega$,

The motion in real space is represented by Figure 2.8(a), Figure 2.8(b) shows the instantaneous state of the particle as given by the representative point in phase space, and Figure 2.8(c) shows the trajectory of the representative point in phase space for rotation with constant angular velocity. The representative point must lie within a band of width 2π in configurational phase space. The angular momentum p_θ may have any positive or negative value. The smooth rotation of the particle is represented by motion from A to B followed by a jump from B to A.

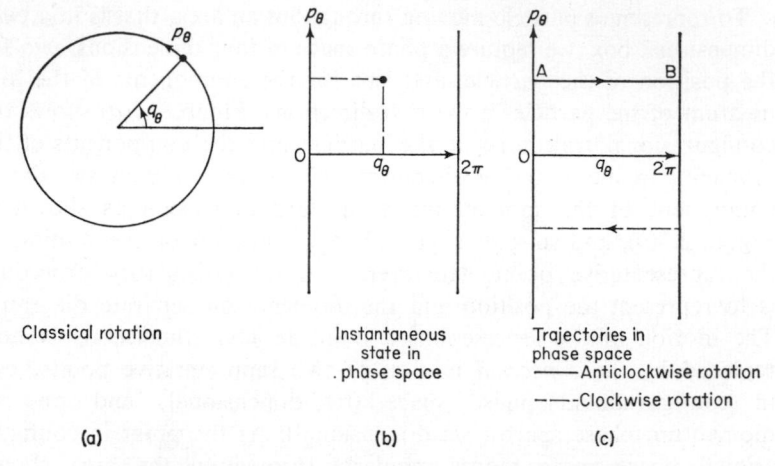

Classical rotation	Instantaneous state in phase space	Trajectories in phase space ——Anticlockwise rotation ––––-Clockwise rotation
(a)	(b)	(c)

Figure 2.8 Phase space for one-dimensional rotation.

(c) For a one-dimensional harmonic oscillator the total energy is constant and is given by

$$\epsilon = \frac{fq^2}{2} + \frac{mv^2}{2} = \frac{fq^2}{2} + \frac{p^2}{2m} \qquad (2.2.1)$$

where f is called the force constant and q is the displacement from the mean position. During the vibrational cycle the potential energy, $fq^2/2$, and the kinetic energy, $mv^2/2$, are continually changing. The variations of potential energy with q and with time are shown in Figure 2.9. If the total energy is ϵ, the motion is restricted to values of q between $+q_\epsilon$ and $-q_\epsilon$ where $q_\epsilon = (2\epsilon/f)^{1/2}$. In order to represent the oscillatory motion as a trajectory in phase space equation (2.2.1) is written in the form

$$\frac{q^2}{2\epsilon/f} + \frac{p^2}{2m\epsilon} = 1 \qquad (2.2.2)$$

This is seen to represent an ellipse whose major and minor axes are respectively $2\sqrt{(2\epsilon/f)}$ and $2\sqrt{(2m\epsilon)}$. The trajectory of constant energy ϵ is therefore an ellipse. Trajectories for different energies will be ellipses one within another as shown in Figure 2.10.

Complex systems have many degrees of freedom for rotation and vibration, and the number of dimensions in phase space required to represent the complete classical motion of an n-particle system is $6n$,

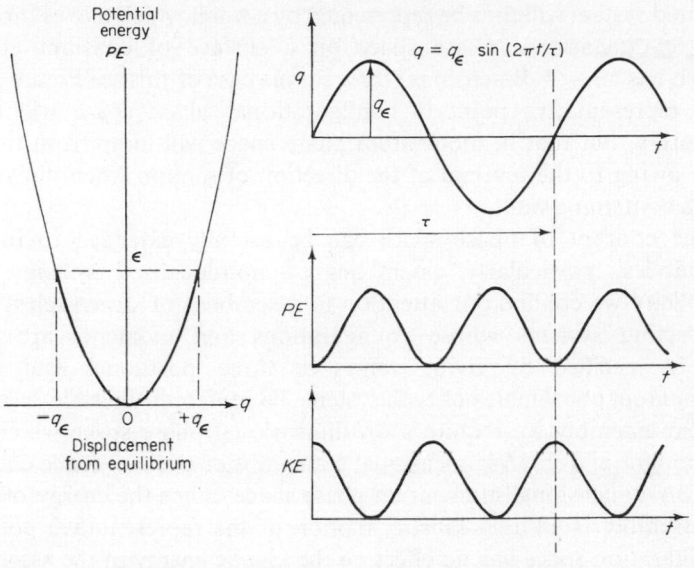

Figure 2.9 Properties of the harmonic oscillator.

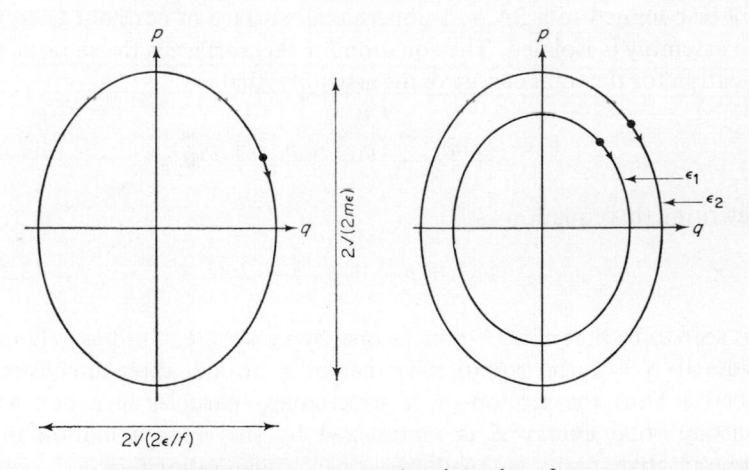

Figure 2.10 Phase space and trajectories of constant energy for the harmonic oscillator.

that is $3n$ dimensions for configurational phase space, and $3n$ dimensions for momentum phase space. The phase space required to represent the motion of a single system is called μ-space after Gibbs. A single isolated system will thus be represented by a point which moves throughout $6n$-dimensional phase space on a surface of constant energy, which has $6n - 1$ dimensions (for a simple case of this see Figure 2.10). The representative point in configurational phase space will move smoothly, but that in momentum phase space will jump from time to time owing to the reversal of the direction of motion when the system hits a containing wall.

The concept of phase space can be usefully extended to include assemblies, particularly assemblies of non-localized systems. For simplicity we confine our attention to assemblies of structureless non-interacting systems whose configurations and momenta are completely specified by giving values to three positional and three momentum coordinates of each system. To represent the state of an N-system assembly we require a $6N$-dimensional phase space which can be thought of as a $3N$-dimensional configurational phase space coupled to a $3N$-dimensional momentum phase space. Since the energy of such an assembly is entirely kinetic, motion of the representative point in configuration space has no effect on the kinetic energy of the assembly. Thus the motion of the representative point in momentum phase space will be confined to a $3N - 1$-dimensional surface of constant energy if the assembly is isolated. The equation for this surface is the same as the equation for the total energy of the assembly, that is

$$E = (1/2m) \sum_{i=1}^{N} (p_{ix}^2 + p_{iy}^2 + p_{iz}^2) \qquad (2.2.3)$$

Rewriting this equation as

$$\sum_i (p_{ix}^2 + p_{iy}^2 + p_{iz}^2) = 2mE \qquad (2.2.4)$$

it is seen to be that of a $3N$-dimensional hyper-sphere of radius $\sqrt{(2mE)}$. (Note if $N = 1$ the equation is that of a normal three-dimensional sphere.) Thus the motion of N structureless particles in a box with constant total energy E is represented by the smooth motion of a representative point is $3N$-dimensional configurational phase space, coupled to the jumpwise motion of a representative point on the surface of a $3N$-dimensional hypersphere in $3N$-dimensional momentum phase space. The jumps occur whenever particles collide with each other or with the walls. The moving points will eventually sample all accessible regions of both configurational and momentum phase space. According to a fundamental postulate of Gibbs, which is equivalent to the fundamental assumption of molecular thermodynamics discussed in Chapter 4, the representative points will, over a sufficient period of time, sample all accessible regions with equal frequency within the limits allowed by the laws of statistics. The phase space appropriate to such an assembly is called γ-space.

2.3   THE CONNECTION BETWEEN CLASSICAL AND QUANTUM MECHANICS

Classical and quantum mechanics represent different models for the behaviour of particles, systems or assemblies. Since the two models give essentially the same results when applied to macroscopic bodies, although their descriptions diverge more and more as the dimensions of the bodies or particles decrease, there ought to be some general procedure for making the transition from the classical to the quantum model. We have already noted that the classical model is essentially more detailed than the quantum model. Thus motion in one-dimension can conveniently be represented by the motion of a representative point in two-dimensional phase space. The position of the point at any instant gives the position and momentum of the particle precisely. According to quantum mechanics it is not possible to know the position and momentum of a particle precisely. The uncertainties in the two quantities are related by the "uncertainty principle" namely

$$\Delta p \times \Delta p \approx h \qquad (2.3.1)$$

The "approximately equals" sign arises since it is not exactly clear what is meant by the uncertainty in the position or the momentum. However, since the quantity $\Delta p.\Delta q$ represents an area in two-dimensional phase space it might be reasonable to suppose that the way of introducing the necessary uncertainty into the classical description of motions in order to obtain an equivalent of the quantum description would be to divide phase space into cells each of an area $\Delta p.\Delta q = h$. Whether the size of the cells should be exactly h or some multiple of h, and whether all the cells ought to be the same size remains to be seen. Although we shall not prove the connection generally we shall examine it for three simple examples.

(a) For a particle in a one-dimensional box of length L, the quantum condition is

$$p = n(h/2L) \qquad (2.3.2)$$

where n may take any positive integral value (Section 2.1). Since the energy of the system is entirely kinetic, $\epsilon = p^2/2m$, lines of constant momentum in the phase plane are also contours of constant total energy. Accordingly the classical phase space trajectory for one-dimensional translation, as we have seen, is a pair of lines parallel to the q-axis and equidistant from it. The line with positive p is traversed from left to right, and the line with negative p from right to left. Two trajectories corresponding to successive quantum states with quantum numbers r and $r + 1$ are shown in Figure 2.11(a). The area between

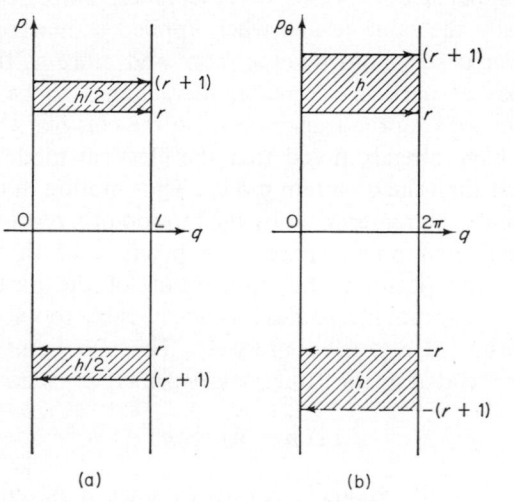

Figure 2.11   The cellular nature of phase space for one-dimensional translation (a) and rotation (b).

these trajectories (shown shaded) is made up of two equal parts whose combined area is

$$\text{Area} = 2 \times L\{(r + 1)(h/2L) - r(h/2L)\} = h \qquad (2.3.3)$$

The complete field of phase space can thus be divided into cells of area h so that the cell boundaries are the classical phase space trajectories corresponding to the quantum mechanical stationary states. Each cell is made up of two equal areas one above and one below the q-axis.

(b) For a one-dimensional rotator the angular momenta allowed by quantum mechanics are obtained from de Broglie's relation (2.1.3), and the quantum condition (2.1.9)

$$p_\theta = I\omega = mvr = j(h/2\pi), \quad j = 0, \pm 1, \pm 2, \ldots \quad (2.3.4)$$

In classical mechanics the angular momentum is a constant of motion, and may be positive or negative depending upon whether the rotation is anticlockwise or clockwise. This places a requirement on the quantum mechanical treatment which is met by allowing j to take all integral values both positive and negative. Each energy level, except that with $j = 0$, is then doubly degenerate.

The classical phase space trajectory for a one-dimensional rotator is a single line parallel to the q_θ-axis, a trajectory above the q_θ-axis representing anticlockwise rotation and one below, clockwise rotation. Two trajectories corresponding to rotational states with successive positive quantum numbers r and $r + 1$ are shown by the full lines in Figure 2.11(b). The area between the trajectories (shown shaded) is $(2\pi)(h/2\pi) = h$; an identical result is obtained if r has a negative value (broken lines). The complete phase plane may again be divided into cells of area h so that each cell boundary represents the classical trajectory corresponding to a quantum mechanical stationary state.

(c) In both examples considered so far the total energy of the system is kinetic, and lines of constant momentum correspond to contours of constant total energy. For the harmonic oscillator this is not so. The trajectories of constant energy in the phase plane are ellipses as shown by equation (2.2.2) whose major and minor axes are $2\sqrt{(2\epsilon/f)}$, and $2\sqrt{(2m\epsilon)}$. The area enclosed by any trajectory is then

$$\text{Area} = (\pi/4) \times (\text{product of major and minor axes})$$

$$= (\pi/4) \times 4 \times 2\epsilon\sqrt{(m/f)}$$

$$= 2\pi\epsilon\sqrt{(m/f)} \quad (2.3.5)$$

In classical mechanics it is shown that the frequency of an harmonic oscillator is

$$\nu = (1/2\pi)\sqrt{(f/m)} \quad (2.3.6)$$

Thus the area enclosed by any trajectory is

$$\text{Area} = \epsilon/\nu \quad (2.3.7)$$

The energy levels allowed by quantum mechanics for the harmonic oscillator are

$$\epsilon = h\nu(n + \tfrac{1}{2}), \qquad \Delta\epsilon = h\nu \qquad (2.3.8)$$

where $\Delta\epsilon$ is the spacing of the energy levels. If trajectories are drawn in the phase plane to represent the energies allowed by quantum mechanics it is readily seen from (2.3.7) and (2.3.8) that the area between successive trajectories is

$$\Delta(\text{area}) = \Delta\epsilon/\nu = h \qquad (2.3.9)$$

This result can be shown to be general and may be stated as follows. For any type of motion in one-dimension the area of phase space between classical trajectories which correspond to successive quantum mechanical stationary states is exactly h. Thus if the precise energies of the quantum states are unimportant one can treat the behaviour of the system semi-classically by assigning one quantum state to each area h in the phase plane.

For motions in s dimensions which require a $2s$-dimensional phase space for their representation, one quantum state corresponds to a cell volume of h^s. Thus if a volume V of phase space is accessible to a system or assembly, the number of quantum states which are accessible is V/h^s.

To be strictly correct a further point must be made. For assemblies of non-localized systems, according to quantum theory, the systems of the assembly are absolutely indistinguishable, and the assembly wave function must be constructed in such a way as to build in this indistinguishability. In the construction of the phase space for an N-particle assembly we assumed that the particles were, in fact, distinguishable by assigning the $6N$ coordinates required to define the phase space. The transition from distinguishability to indistinguishability of systems is readily made, for there would in fact be $N!$ equivalent points in phase space for every real state of the assembly if the systems were unlabelled. These would represent distinct states if they were labelled. We thus obtain a somewhat modified result.

For assemblies of localized systems number of quantum states corresponding to a volume V of phase space $= V/h^s$ $\qquad$ (2.3.10)

For assemblies of non-localized systems, number of quantum states corresponding to a volume V of phase space $= V/(h^s N!)$ $\quad$ (2.3.11)

It might appear from (2.3.10) and (2.3.11) that the number of states accessible to an assembly of localized systems must be vastly greater than the number of states accessible to an assembly of non-localized systems because $N!$ is so large a number for any real assembly. This is not in fact the case because the volume of configurational phase space accessible to an assembly of localized systems is so much smaller than that accessible to an assembly of non-localized systems. This difference normally more than cancels out the effect of the $N!$

PROBLEMS

2.1 What is the wavelength of an electron whose kinetic energy is 100 electron volts? Does the result have any relevance to the possible resolving power of an electron microscope? Would the resolving power in theory be improved by using protons instead of electrons?

2.2 To a rough approximation the electrons in benzene may be thought of as being constrained to move in a circle of radius $1 \cdot 2 \times 10^{-10}$ m. Using conditions (2.1.9) and (2.1.10) calculate the difference in energy $\Delta\epsilon$ between states for which $j = 0$ and $j = 1$. Using the quantum condition $\Delta\epsilon = h\nu$ find the frequency and wavelength of radiation which would excite this transition. The absorption maximum for benzene occurs at about 250 nm. Comment on any connection.

2.3 Calculate the quantum state density for Ar atoms contained in a length of 1 m with an energy of $\frac{1}{2}kT$. Use equation (2.1.7) to obtain $dn/d\epsilon$, that is the value of $N(\epsilon)$. What are the units? What is the value of n (the translational quantum number) for an atom with $\epsilon = \frac{1}{2}kT$ in a length of 1 m? Why is $N(\epsilon)$ so much larger than n?

2.4 If a single particle in a box is in a stationary state, why can no meaning be attached to the idea that it has a particular position in the box at any instant?

2.5 Calculate the translational quantum number for one dimension for a nitrogen molecule whose kinetic energy is $\frac{1}{2}kT$ and is contained in a box of side 10^{-2} m at 300 K.

Calculate the rotational quantum number for a nitrogen molecule with an energy kT at 300 K given that the internuclear distance is $1 \cdot 10 \times 10^{-10}$ m.

CHAPTER 3

CLASSICAL THERMODYNAMICS

3.1 THE LAWS OF THERMODYNAMICS

Classical thermodynamics is based upon four laws of experience to which no exceptions have been found or are ever likely to be found. The laws make no mention of the molecular concept. They may be stated as follows:

Zeroth Law: If two bodies are independently in thermal equilibrium with a third, they are in thermal equilibrium with each other. Alternatively: the degree of hotness (or temperature) is a measurable property of any body internally at equilibrium.

First Law: The internal energy of a completely isolated assembly is invariant with respect to time.

Second Law: Natural processes never occur exactly in reverse. When they occur within an isolated assembly they are associated with an increase in the value of a property which is called the entropy of the assembly. The definition of entropy requires the collateral definition of a thermodynamic scale of temperature.

Third Law: The entropy change in any reaction between perfect crystals at a thermodynamic temperature of zero is zero.

These laws are of a most general nature. The first and second laws were formulated during the nineteenth century and are associated with the names of Joule, Carnot and Clausius. The third law was discovered through the work and insight of Nernst, Planck and others. The zeroth law has been introduced only recently, having been unwittingly assumed in earlier treatments of thermodynamics. It completes the logical foundations of classical thermodynamics.

The zeroth law is equivalent to the statement that the hotness of any body can, in principle, always be measured by some independent body (for example, a thermometer), provided only that some property of the reference body changes continuously and monotonically with the degree of hotness. There are many suitable properties, for example, the resistance of a piece of wire, the length of a copper rod, the pressure of a fixed mass of gas in a constant volume, the volume of a fixed mass of liquid under a constant pressure, the viscosity of a liquid, the vapour pressure of a pure liquid. The law defines the idea of hotness or temperature, but not the scale of temperature. Indeed it emphasises that there are innumerable ways of measuring temperature.

The first law, concerned with the conservation of energy, defines the internal energy, U, of an assembly. The internal energy of a completely isolated assembly is constant, but that of an assembly in contact with its surroundings may be changed. If the interaction is through a thermally conducting wall, the change in internal energy, ΔU, is equal to the quantity of heat which passes into the assembly. If the interaction is via a wall which transmits pressure, ΔU is equal to the work done on the assembly by its surroundings. We thus obtain the familiar equation embodying the first law:

$$U = q + w$$

$$= \text{heat absorbed by assembly} + \text{work done on assembly} \tag{3.1.1}$$

Although there are other ways of changing U, for example by absorption of radiation or electrical energy, such changes can usually be classified by their effects as equivalent either to heat or work, and so are not specifically included in (3.1.1).

When the second law in the general form "Natural processes never occur exactly in reverse" is applied to certain cyclic processes called Carnot cycles (cycles which involve taking a working fluid through two pairs of alternate isothermal and adiabatic stages which eventually return it to its original state), a quantity emerges which can be shown to be a property of all thermodynamic assemblies. This property is called the entropy. The difference in the entropy of any assembly between two states a and b is found to be

$$\Delta S = S_b - S_a = \int_a^b \frac{dq_{rev}}{\theta} \tag{3.1.2}$$

dq_{rev} is the heat absorbed in an infinitesimal part of any reversible path connecting the states a and b, and θ is the thermodynamic temperature of the assembly. The general treatment shows that the thermodynamic scale of temperature is unique apart from the size of the temperature unit, but does not define it in practical terms. To establish this definition it is necessary to examine a Carnot cycle which uses a working fluid with a known equation of state. The best fluid to use is the ideal gas since it then transpires that the thermodynamic scale of temperature is identical (again apart from the arbitrary choice of the size of the unit) to the ideal gas scale of temperature. By convention the ideal gas and thermodynamic scales are defined by equation

$$T = PV/A \quad \text{(ideal gas)} \tag{3.1.3}$$

where P and V are the pressure and volume of an arbitrary mass of an ideal gas, and A is chosen so that $T = 273 \cdot 15$ exactly when the gas is immersed in an equilibrium mixture of pure ice, liquid water and water vapour. The temperature scale so derived is called the absolute scale, and the unit is the Kelvin. The normal freezing point of water under a pressure of 1 standard atmosphere ($1.01325 \times 10^5 \text{ N m}^{-2}$) is then $273 \cdot 16$ K. The normal boiling point at the same pressure is $373 \cdot 16$ K.

Entropy changes are thus measured in practice by application of equation (3.1.4)

$$S = S_b - S_a = \int_a^b \frac{\mathrm{d}q_{\mathrm{rev}}}{T} \tag{3.1.4}$$

Having defined entropy by (3.1.4), it is readily confirmed that "natural processes within an isolated assembly are always accompanied by an increase in the entropy". Only if a reversible process occurs does the entropy of an isolated assembly remain constant.

The statement of the second law thus requires two sub-statements:

(a) There is a unique thermodynamic scale of temperature which is identical with the ideal gas scale, and

(b) there exists a property of all assemblies called the entropy. For an isolated assembly the entropy can never decrease. The entropy change is given by equation (3.1.4)

The second law taken in conjunction with the zeroth and first laws enables thermodynamic relationships for assemblies in equilibrium to be derived.

For practical chemical purposes the third law is equivalent to the statement that the entropy of all perfect crystals at the absolute zero of temperature is zero. It re-emphasizes that an absolute zero of temperature exists, and asserts that it is permissible to take an absolute zero of entropy. In this respect entropy differs from internal energy and the derived free energy functions, for which no ultimate zeros exist.

3.2 THERMODYNAMIC RELATIONSHIPS

The laws of thermodynamics give rise to a number of general equations which are of great practical value.

According to the first law the change in internal energy when an infinitesimal reversible change takes place in an assembly of fixed material content is given by

$$dU = dq_{rev} + dw_{rev} \tag{3.2.1}$$

The heat absorbed in a reversible process is given by the second law as

$$dq_{rev} = T \, dS \tag{3.2.2}$$

and the work done on the assembly via a wall which can transmit pressure is

$$dw_{rev} = - p \, dV \tag{3.2.3}$$

where V is the volume of the assembly. Reversibility in regard to work implies that the internal and external pressures differ only by infinitesimal amounts. We thus obtain generally

$$dU = T \, dS - p \, dV \tag{3.2.4}$$

The restriction to fixed material content does not, of course, rule out chemical or phase changes within the assembly for these do not alter quantities of elements present.

While equation (3.2.4) embodies much of the first and second laws it is convenient in practice to define three other thermodynamic quantities, the enthalpy, H, the Helmholtz free energy, F, and the Gibbs free energy, G.

$$H \equiv U + pV \tag{3.2.5}$$

$$F \equiv U - TS \tag{3.2.6}$$

$$G \equiv H - TS \equiv U + pV - TS \tag{3.2.7}$$

Differentiation and substitution of 3.2.4 then yields

$$dH = T \, dS + V \, dp \tag{3.2.8}$$

$$dF = -S\,dT - p\,dV \tag{3.2.9}$$

$$dG = -S\,dT + V\,dp \tag{3.2.10}$$

From these we obtain alternative definitions of entropy

$$-S = \left(\frac{\partial F}{\partial T}\right)_V = \left(\frac{\partial G}{\partial T}\right)_p \tag{3.2.11}$$

Incorporating these equations into (3.2.9) and (3.2.10) yields the Gibbs–Helmholtz equations (3.2.12) and (3.2.13)

$$\left\{\frac{\partial(F/T)}{\partial T}\right\}_V = -\frac{U}{T^2} \tag{3.2.12}$$

$$\left\{\frac{\partial(G/T)}{\partial T}\right\}_p = -\frac{H}{T^2} \tag{3.2.13}$$

When a reversible process is carried out at constant temperature or at constant temperature and pressure equations (3.2.9) and (3.2.10) give

$$dF_T = -p\,dV = dw_{rev} \tag{3.2.14}$$

$$dG_{p,T} = 0 \tag{3.2.15}$$

The second of these expressions is widely used in thermodynamics to derive practical equations for chemical and phase equilibria.

The condition for equilibrium between two phases of the same pure component may be derived directly. Suppose an ampoule containing gaseous and liquid benzene is placed in a thermostat and allowed to come to equilibrium. The equilibrium is characterized by a temperature, T, and an equilibrium vapour pressure, p. Suppose that δn moles of benzene evaporate into the vapour phase, the quantity δn being so small that the process may be regarded as reversible. Since the process is carried out at constant pressure and temperature, equation (3.2.15) applies. If $G_{m(v)}$ and $G_{m(l)}$ are the molar Gibbs free energies of the vapour and liquid then the equilibrium condition is

$$0 = G = \delta n\{G_{m(v)} - G_{m(l)}\} \tag{3.2.16}$$

and since δn is not zero

$$G_{m(v)} = G_{m(l)} \tag{3.2.17}$$

Since this equality must hold at all temperatures, the vapour pressure must adjust itself with temperature so that (3.2.17) holds. To obtain a practical expression for p as a function of T equation (3.2.10) is used to eliminate G. Thus for any change in temperature of the thermostat, dT,

$$dG_{m(v)} = dG_{m(1)} \quad \text{or} \quad -S_{m(v)} \, dT + V_{m(v)} \, dp = -S_{m(1)} \, dT + V_{m(1)} \, dp$$

or

$$dp/dT = \Delta S_m/\Delta V_m \qquad (3.2.18)$$

where ΔS_m and ΔV_m are the differences in molar entropy and molar volume between the two phases. For any phase change

$$\Delta S_m = L_m/T \qquad (3.2.19)$$

where L_m is the latent heat of transition. The final equation for a single component assembly is therefore

$$dp/dT = L_m/T\Delta V_m \qquad (3.2.20)$$

Equation (3.2.20) applies to any type of phase transition, solid to solid, solid to liquid, solid to vapour or liquid to vapour. When one of the phases is a dilute vapour the volume of the condensed phase may to a first approximation be ignored in ΔV_m, and $V_{m(v)}$ may be set equal to RT/p. Replacing ΔV_m by RT/p then yields

$$d \ln p/dT = L/RT^2 \quad \text{(dilute vapour)} \qquad (3.2.21)$$

which is the well known Clapeyron–Clausius equation.

On turning to multicomponent assemblies the equilibrium condition can no longer be stated in terms of molar free energies of pure substances, it is necessary to work in terms of partial molar free energies, or chemical potentials. The chemical potential of a component in an assembly is best introduced and defined by considering an open assembly, that is one whose material content may be altered by addition of material from outside. If dn_1 moles of component 1, dn_2 moles of component 2, etc. are added reversibly to an assembly, changes in internal energy, enthalpy, Helmholtz free energy and Gibbs free energy result from the extra material added. They are given by equations (3.2.22) to (3.2.25).

$$dU = T \, dS - p \, dV + \sum \mu_i \, dn_i \qquad (3.2.22)$$

$$dH = T \, dS + V \, dp + \sum \mu_i \, dn_i \qquad (3.2.23)$$

$$dF = -S \, dT - p \, dV + \sum \mu_i \, dn_i \qquad (3.2.24)$$

$$dG = -S \, dT + V \, dp + \sum \mu_i \, dn_i \qquad (3.2.25)$$

where the μ_i are the chemical potentials of the different components. Of these equations only the last two are of practical interest as is seen by writing the equations defining the μ_i

$$\mu_i = (\partial U/\partial n_i)_{S,V,n_1,n_2,\ldots}$$
$$= (\partial H/\partial n_i)_{S,p,n_1,n_2,\ldots}$$
$$= (\partial F/\partial n_i)_{T,v,n_1,n_2,\ldots}$$
$$= (\partial G/\partial n_i)_{T,p,n_1,n_2,\ldots} \tag{3.2.26}$$

It is clearly impossible in practice to keep the entropy constant when material is added to an assembly since the material added will bring its own entropy with it, and the magnitude of this additional entropy cannot readily be calculated, although, in principle, it can be found using the third law. The last definition involving the Gibbs free energy is the most satisfactory since it shows the chemical potential to be identical with the partial molar Gibbs free energy. The meaning of the chemical potential (and indeed by similar arguments of any partial molar quantity) may be clarified if we imagine the assembly to be increased in size at constant temperature and pressure by adding to it reversibly small amounts of each constituent in proportions *which keep the overall composition constant*. The change in G is then

$$dG_{T,p} = \sum \mu_i \, dn_i \tag{3.2.27}$$

If the assembly initially contained n_1 moles of component 1, n_2 moles of 2 and so on, then the dn_i are by definition in the same ratio as the n_1, and we can write $dn_i = n_i dx$, where dx is now the fractional increase in the size of the assembly resulting from one round of small additions. Thus

$$dG_{T,p} = dx \sum \mu_i n_i \tag{3.2.28}$$

This equation may be integrated from $x = 0$ to $x = 1$ giving

$$G = \sum \mu_i n_i \tag{3.2.29}$$

The total free energy of an assembly is therefore the sum over all components of the product of the chemical potential and the number of moles.

When an assembly contains several phases each of several components the total free energy is

$$G = \sum_j \sum_i \mu_{ij} n_{ij} \tag{3.2.30}$$

where the subscript "ij" refers to the ith component of the jth phase. A slight modification of the argument used for a single component shows that the condition for phase equilibrium is

$$u_{ij} = \mu_{ik} \tag{3.2.31}$$

For assemblies in equilibrium equation (3.2.30) thus reduces to

$$G = \sum_j \sum_i \mu_i n_{ij} \tag{3.2.32}$$

The condition for chemical equilibrium is also readily obtained. Consider the reaction

$$a.A + b.B = c.C$$

Suppose that A, B and C are sealed in an ampoule and placed in a thermostat until chemical equilibrium is established. Now let $a.dx$ moles of A react with $b.dx$ moles of B to produce $c.dx$ moles of C where dx is an infinitesimal fraction of the total quantity of material in the assembly. The assembly as a whole is closed and has suffered an infinitesimal reversible process at constant temperature and pressure. The change of Gibbs free energy is then zero

$$0 = dG = \sum \mu_i \, dn_i$$
$$= -a.dx.\mu_a - b.dx.\mu_b + c.dx.\mu_o$$
$$= dx(-a.\mu_a - b.\mu_b + c.\mu_c) \tag{3.2.33}$$

Since dx is not zero we obtain, inserting the superscript ^{e} to denote equilibrium,

$$a.\mu_a{}^e + b.\mu_b{}^e = c.\mu_c{}^e, \quad \text{or} \quad \Delta G^e = 0 \tag{3.2.34}$$

This equation can obviously be extended to any number of reactants and products, and can be stated in the form:

{Total chemical potential of reactants}

= {total chemical potential of products at equilibrium}   (3.2.35)

For reactions involving ideal gases the chemical potentials of the different substances are the same as their molar Gibbs free energies would be if they were in the pure state at the same concentration. The chemical potential may thus be expressed for any temperature as

$$\mu = \mu^* + RT \ln (p/p^*) \tag{3.2.36}$$

where μ^* is the chemical potential at a standard pressure p^*. When the variable pressure p is an equilibrium partial pressure p^e we obtain

$$\mu^e = \mu^* + RT \ln (p^e/p^*) \tag{3.2.37}$$

Replacing the equilibrium chemical potentials in equation (3.2.34) by their values according to equation (3.2.37) gives after some rearrangement:

$$\Delta G^* \equiv c\mu_c^* - a\mu_a^* - b\mu_b^*$$

$$= RT \ln \frac{(p_c^*)^c}{(p_a^*)^a (p_b^*)^b} - RT \ln \frac{(p_c^e)^c}{(p_a^e)^a (p_b^e)^b}$$

$$= RT \ln Q_p^* - RT \ln K_p \tag{3.2.38}$$

In equation (3.2.38), K_p is the equilibrium constant for the reaction written in terms of pressures, while Q_p^* is a quotient constructed in the same way as K_p but using the standard pressures in place of the equilibrium pressures. ΔG^* is the change in Gibbs free energy when the reaction takes place between reactants at standard pressures to give products at standard pressures. Equation (3.2.38) is known as the Van't Hoff Isotherm. When the standard pressures are taken as unity, then

$$\Delta G^\circ = -RT \ln K_p \tag{3.2.39}$$

When (3.2.38) is inserted into the Gibbs–Helmholtz equation (3.2.13) the Van't Hoff Isochore results

$$\left(\frac{\partial \ln K_p}{\partial T}\right)_p = \frac{\Delta H^\circ}{RT^2} \tag{3.2.40}$$

The condition of constancy of pressure in equation (3.2.40) is somewhat misleading and is indeed unnecessary. It is already met by writing the isotherm in the form (3.2.38) since constancy of pressure simply implies constancy of the initial and final pressures, that is constancy of Q_p^*. But Q_p^* is a constant by definition.

The third law, in principle, enables S° to be obtained by purely calorimetric measurements made on the reactants and products individually. Since $S_0^\circ = 0$ at 0 K, the entropy of any substance is obtained from the general formula

$$S_T^\circ - S_0^\circ = \sum \int_{T_{ij}}^{T_{jk}} C_{p(j)} \, d \ln T + \sum \frac{L_{ij}}{T_{ij}} \tag{3.2.41}$$

where $C_{p(j)}$ is the molar heat capacity of phase j which exists between the temperatures T_{ij} and T_{jk}, and L_{ij} is the latent heat for phase transition between phase i and phase j.

Its practical value in chemistry is twofold. Firstly since $\Delta G^\circ = -RT \ln K_p$, and since $\Delta G^\circ = \Delta H^\circ - T\Delta S^\circ$, one can obtain equilibrium constants for chemical reactions from purely calorimetric measurements. Alternatively when ΔH° is unknown it may sometimes be possible to obtain it if a value for K_p is available. An example of the accurate determination of ΔH° from a knowledge of K_p and calculated values of ΔS° is given in Chapter 11 for the reaction $F_2 = 2\,F$.

CHAPTER 4

FOUNDATIONS OF MOLECULAR THERMODYNAMICS

4.1 THE ISOLATED ASSEMBLY, AND THE FUNDAMENTAL ASSUMPTION OF MOLECULAR THERMODYNAMICS

If an assembly of systems is completely isolated from the outside world, its material content and total energy must remain constant for all time. In the present treatment of the isolated assembly we assume also that the assembly has a fixed shape and volume, as it would have, for instance, if it were enclosed within a rigid wall, which itself could be regarded as part of the assembly.

According to wave mechanics it is possible, in principle if not in practice, to specify all the quantum states and associated energies for any assembly of defined constitution, volume and shape. For a completely isolated assembly only those states whose energies have the precise value E will be accessible, although any slight relaxation of the isolation will make other states accessible. For assemblies of practical importance in chemistry (say those containing more than 10^{10} systems) the density of quantum states around any likely energy E is astronomical and the actual number of quantum states with an energy of precisely E is likely to be exceedingly large. Any isolated assembly of practical size can therefore exist in a very large number of quantum states. Each of these quantum states corresponds to a distinct complexion of the assembly. We might therefore ask "Which of this large number of states or complexions is most likely to represent the actual state of the assembly at any instant? Are there any states or complexions which are especially favoured relative to the others, and if so, why?"

As far as it has been possible to discover there is no preference whatever for any particular quantum state or complexion, and it is not possible to conceive of any reason why certain quantum states or complexions should be more likely than others. The fundamental assumption asserts this belief as a fact, and it may be stated:

ALL ACCESSIBLE QUANTUM STATES OF AN ISOLATED ASSEMBLY, WHICH NECESSARILY HAVE THE SAME ENERGY, ARE EQUALLY PROBABLE. THE INSTANTANEOUS STATE OF AN ISOLATED ASSEMBLY IS EQUALLY LIKELY TO BE ANY OF THE ACCESSIBLE STATES.

If there were indeed certain amongst the accessible states which were to be preferred, they would have to be distinguishable by some property of the states which could indicate their degree of superiority. However, the only external properties or constraints which are recognized in the construction of the wave functions for the assembly are the total energy, volume, shape and constitution of the assembly. Although these properties strongly affect the forms of the wave functions, no other *independent* properties can be extracted from them. There is therefore no conceivable reason for bias in favour of any one quantum state or group of quantum states.

The fundamental assumption may also be stated in terms of the ensemble of accessible states, which for an isolated assembly is the micro-canonical ensemble. According to Gibbs, the long-term average of any property of a real assembly is identical to the properly weighted mean taken over the appropriate ensemble. The fundamental assumption defines the method of weighting for the micro-canonical ensemble:

EACH MEMBER OF THE MICRO-CANONICAL ENSEMBLE HAS EQUAL WEIGHT IN DETERMINING ANY AVERAGE PROPERTY, AND HAS AN EQUAL PROBABILITY OF REPRESENTING THE INSTANTANEOUS STATE OF A REAL ISOLATED ASSEMBLY.

We have shown above that the quantum description of systems or assemblies can be connected with the classical description by introducing a cellular structure into phase space. Accordingly it is possible to express the fundamental assumption in terms of phase space rather than in terms of quantum states. This was indeed the way in which it was originally expressed by Gibbs. The classical description of the evolution of an assembly with time involves the movement of the representative point through phase space. For an assembly of N structureless systems, a phase space of $6N$ dimensions is required, comprising $3N$ dimensions in configurational phase space, and $3N$ in momentum phase space. In the same way as a single particle, in a rough walled box, samples all regions of the box in the long term with equal frequency, so the representative point for an assembly of non-interacting structureless systems samples all regions of configurational phase space with equal frequency. If the assembly is isolated and hence its energy constant, the representative point in momentum phase space must lie on the surface of a $3N$-dimensional sphere of radius $(2mE)^{1/2}$ where E is the total energy, and m the mass of a system (equation 2.2.4). As the assembly develops in time the representative point in configurational phase space moves smoothly, while that in momentum phase space jumps from point to point in an apparently random way. It is reasonable to suppose that over a long period all regions on the surface of the sphere will be visited equally frequently. This view is equivalent to the fundamental assumption which may be stated in the alternative form:

THE REPRESENTATIVE POINT FOR AN ISOLATED ASSEMBLY IS EQUALLY LIKELY TO BE FOUND IN ANY ACCESSIBLE REGION OF PHASE SPACE, VOLUME FOR VOLUME.

We may complete the correlation between the quantum and classical descriptions by constructing a classical ensemble. This is done by imagining that points are scattered with equal density throughout the whole of phase space. We then assign one member of the ensemble to each such point. The micro-canonical ensemble contains members only for those points which lie on the surface of a sphere in momentum phase space of radius $(2mE)^{1/2}$. The statement of the fundamental assumption in terms of the classical ensemble is then identical to the statement in terms of the quantum ensemble.

4.2 THE ASSEMBLY IN A HEAT BATH, AND THE ASSEMBLY PARTITION FUNCTION

An assembly of fixed constitution, shape and volume in a heat or energy bath will nearly always have an energy close to the long term average $\bar{E}$, which we can identify with the thermodynamic internal energy, U, but from time to time small fluctuations, and very occasionally larger fluctuations, will occur due to the random motion of the molecules. There is thus a finite possibility that the assembly will be found in a quantum state whose energy differs appreciably from U. If the assembly is sufficiently small (for example a few molecules or even a single molecule) the fluctuations from the mean may be very large. The key to molecular thermodynamics is the evaluation of this probability. The evaluation can be carried out with *no further assumption* than that stated in Section 4.1.

We proceed by noting first that the fundamental assumption will apply to states of a specified energy, E, whether the assembly is isolated or not. This leads to the important corollary that the probability that a given thermostated assembly is found in a state of energy E, depends only upon E and the communicable characteristic of the thermostat, namely its temperature. For an assembly in a thermostat the fundamental assumption may be stated in the slightly more general form:

ALL ACCESSIBLE QUANTUM STATES OF AN ASSEMBLY IN A HEAT BATH WHICH HAVE AN ENERGY E ARE EQUALLY PROBABLE. THE VALUE OF THIS PROBABILITY FOR A SPECIFIED ASSEMBLY DEPENDS ONLY UPON E AND THE TEMPERATURE OF THE BATH.

This form of the fundamental assumption is the starting point for the main derivation which leads to the establishment of equations for the thermodynamic parameters of an assembly. In what follows we denote the probability that an assembly is found in a specified quantum state, i, whose energy is E_i, by the symbol $P(\text{state}, i)$ or, if we wish to draw attention to the energy of the state, by the symbol $P(\text{state}, E_i)$. We shall denote the probability that the assembly has an energy E_i, irrespective of the particular quantum state by the symbol $P(\text{energy}, E_i)$. Thus by the fundamental assumption

$$P(\text{energy}, E_i) = \Omega_i P(\text{state}, E_i) \qquad (4.2.1)$$

where Ω_i is the degeneracy of the energy level E_i, that is, the number of quantum states with an energy E_i.

If we wish to use the notation of quantum state density rather than degeneracy we write

$$P'(\text{energy}, E)\, dE = N(E) P(\text{state}, E)\, dE \qquad (4.2.2)$$

$P'(\text{energy}, E)\, dE$ is then the probability that the assembly exists in any state whose energy lies between E and $E + dE$. $P'(\text{energy}, E)$ is called the occupation probability per unit energy range at an energy E. It has the same units as $N(E)$ the quantum state density, namely (energy)$^{-1}$.

When we wish to refer to a particular assembly, say assembly k, we shall use subscripts on the P. For example $P_k(\text{state } i)$ is the probability that assembly k is in state i.

To evaluate the probability that a thermostatted assembly is in a quantum state i of energy E_i, we imagine a global assembly (not to be confused with an ensemble) which contains an arbitrary number L of sub-assemblies. At first we shall suppose that the L sub-assemblies are of identical constitution, shape and volume, but we remove this restriction later. The sub-assemblies are regarded as being in weak thermal contact and so able to possess a range of energies, but the contact is sufficiently weak that each sub-assembly has its own private set of quantum states with well defined energies. These energies are not affected by the states of the neighbouring sub-assemblies. The global assembly, on the other hand, is taken as isolated and to have a precise energy E_{global}.

Although it is unnecessary to specify the model further, it might be helpful to imagine the global assembly as composed, say, of one thousand 1 cm diameter spheres of sodium chloride arranged in a cubic array 10 cm across. A section through such an assembly is shown in Figure 4.1.

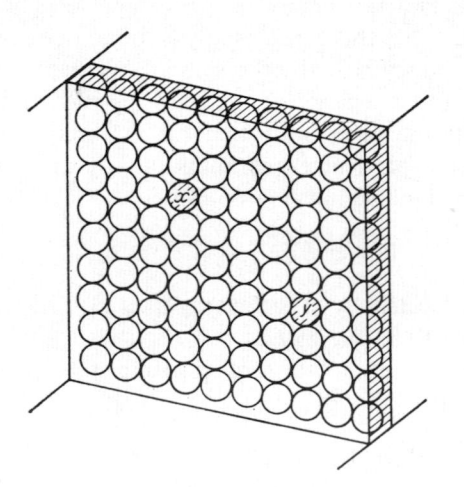

Figure 4.1 Section through global assembly containing 100 sub-assemblies.

According to quantum mechanics there will be definable quantum states, energies and wave functions for each sub-assembly, and there will be corresponding quantum states for the global assembly. In a sense the sub-assemblies may be thought of as the "systems" of the global assembly. Thus the wave functions and energies of the global and sub-assemblies are related by equations (4.2.3) and (4.2.4)

$$\Psi_{\text{global}} = \Psi_a(1)\, \Psi_b(2)\, \Psi_c(3)\ldots\Psi_n\,(L) \qquad (4.2.3)$$

$$E_{\text{global}} = E_1 + E_2 + \ldots + E_L \qquad (4.2.4)$$

where $\Psi_i(j)$ is the wave functions for the jth assemblies in state i, there being L factors in Ψ_{global}.

Consider now two sub-assemblies x and y which are *well separated* in the global assembly (see Figure 4.1). Suppose that they are in quantum states of energies $E_{x(1)}$ and $E_{y(1)}$, that their total energy is E_{xy}, that is

$$E_{xy} = E_{x(1)} + E_{y(1)} \qquad (4.2.5)$$

and let the remainder of the sub-assemblies be in particular defined quantum states. Now suppose that another pair of quantum states of x and y exist whose total energy is also E_{xy}, but where the individual energies are $E_{x(2)}$ and $E_{y(2)}$. This pair can be combined with the same combination of states for the other $(L - 2)$ sub-assemblies as before. We have thus defined two quantum states of the global assembly. By the fundamental assumption they are equally probable. Thus, since the $(L - 2)$ sub-assemblies are in the same quantum states for the two different combinations of the states of x and y we can write

$$\left\{ \begin{matrix} \text{Probability that } x \text{ is in state of energy } E_{x(1)}, \\ \text{while } y \text{ is in a state of energy } E_{y(1)} = E_{xy} - E_{x(1)} \end{matrix} \right\}$$

$$= \left\{ \begin{matrix} \text{Probability that } x \text{ is in a state of energy } E_{x(2)} \\ \text{while } y \text{ is in a state of energy } E_{y(2)} = E_{xy} - E_{x(2)} \end{matrix} \right\} \qquad (4.2.6)$$

Since x and y are widely separated there is no reason why there should be any correlation between their quantum states. Thus the value of $P_x(\text{state}, E_{x(1)})$ cannot be affected by the quantum state occupied by y. The probabilities P_x and P_y are independent. By the laws of probability the probability that two independent events occur in conjunction is the product of the probabilities that they occur singly. Thus

$$P_x(\text{state}, E_{x(1)}) \times P_y(\text{state}, E_{y(1)}) = P_x(\text{state}, E_{x(2)}) \times P_y(\text{state}, E_{y(2)}) \qquad (4.2.7)$$

Now if the local assemblies x and y are identical in constitution, the subscripts x and y can be removed, giving

$$P(\text{state}, E_1) \times P(\text{state}, E - E_1) = P(\text{state}, E_2) \times P(\text{state}, E\text{–}E_2) \qquad (4.2.8)$$

Since there are no restrictions on E_1 or E_2 except that both must be less than E, which itself is only restricted to E_{global}, we can generalize (4.2.8) to give

$$P(\text{state, } E_i) \times P(\text{state, } E–E_i) = (\text{constant dependent only on } E)$$
$$(4.2.9)$$

The only known function which fits this equation for all E_i is

$$P(\text{state, } E_i) = (1/Q) \exp{[-\beta E_i]} \qquad (4.2.10)$$

where Q and β are constants which, from a purely mathematical viewpoint, can be chosen freely. It is readily seen by substitution that (4.2.10) satisfies (4.2.9)

$$(1/Q) \exp{[-\beta E_i]} \times (1/Q) \exp{[-\beta(E–E_i)]} = (1/Q^2) \exp{[-\beta E]}$$

$$= (\text{constant dependent only upon } E) \qquad (4.2.11)$$

Suppose we now carry out the derivation for sub-assemblies x and y which have different constitutions. The argument proceeds as before until equation (4.2.7), but thereafter the suffixes must be retained. Equation (4.2.9) now becomes

$$P_x(\text{state, } E_i) \times P_y(\text{state, } (E–E_i)) = \text{constant} \qquad (4.2.12)$$

The only functions which satisfy this condition irrespective of E_i are

$$P_x(\text{state, } E_i) = (1/Q_x) \exp{[-\beta E_i]}$$
$$P_y(\text{state, } E–E_i) = (1/Q_y) \exp{[-\beta(E–E_i)]} \qquad (4.2.13)$$

which, when inserted into (4.2.12) give

$$P_x \times P_y = 1/(Q_x Q_y) \exp{[-\beta E]}$$

$$= (\text{constant dependent only upon } E) \quad (4.2.14)$$

As before Q_x, Q_y and β may be chosen at will. In order to interpret them we first notice that the Q's may be different for different sub-assemblies, and so may be functions of the constitutions of the assemblies, whereas β must be the same for all sub-assemblies, for assembly x may be paired with any other sub-assembly in the global assembly. Secondly, from equation (4.2.14), the probability of finding two sub-assemblies x and y in a specified pair of quantum states of total energy E is again composed of a "Q-factor" and an exponential factor. The Q-factor is the product of the Q's for the two constituent assemblies, but the constant β remains unchanged being the same for the pair as for the individuals. By combining sub-assemblies in larger groups it is clear that β has the same value for a group of any size, and is indeed the same as for the global assembly, whereas Q for a group is the product of the Q's for the constituent assemblies. β is therefore an *intensive* property of the global assembly and of the sub-assemblies while ln Q is an *extensive* property of any assembly. (Note: An intensive property is one whose value is independent of the size of an assembly, for instance pressure, temperature, density; an extensive property is one whose value is proportional to the size of the assembly, for instance, mass, internal energy, volume, entropy.)

If we now think of the global assembly as the heat bath, and a particular sub-assembly as the "assembly in a heat bath" we notice that β is the property which is common to both. There is therefore a strong inclination to identify β with some measure of temperature. We shall prove this identification shortly, but first it is necessary to consider the interpretation and meaning of Q.

Since any assembly must exist in some quantum state, the sum of the probabilities over all accessible states must be unity, Thus

$$1 = \sum_{\text{states}} P(\text{state}, i)$$

$$= P(\text{state}, 1) + P(\text{state}, 2) + \ldots$$

$$= (1/Q) \exp[-\beta E_1] + (1/Q) \exp[-\beta E_2] + \ldots$$

$$= (1/Q) \{\exp[-\beta E_1] + \exp[-\beta E_2] + \ldots\} \quad (4.2.15)$$

Whence

$$Q = \exp[-\beta E_1] + \exp[-\beta E_2] + \ldots$$

$$= \sum_{\text{states}} \exp[-\beta E_i] \quad (4.2.16)$$

This value for Q is forced by the meaning of the probability that a quantum state is occupied. The acceptance of a specific value of Q is perfectly in order since the mathematical argument starting just before equation (4.2.5) allows any value to be chosen which happens to be convenient. Q as defined by equation (4.2.16) is called the SUM OVER STATES, or more commonly the PARTITION FUNCTION FOR THE ASSEMBLY. The partition function is the key function in molecular thermodynamics and values for all thermodynamic parameters may be derived from it.

The probability of finding an assembly in a given state is now written:

$$P(\text{state, } i) = \frac{\exp[-\beta E_i]}{\sum\limits_{\text{states}} \exp[-\beta E_i]}$$

$$= \exp[-\beta E_i]/Q \qquad (4.2.17)$$

Often it is convenient to sum over energy levels rather than over distinguishable quantum states. We nevertheless still have to ensure that Q contains in effect one term of the form $\exp[-\beta E_i]$ for every state. Thus if the degeneracy of an energy level E_i is Ω_i, the term $\exp[-\beta E_i]$ will have to appear Ω_i times in the summation for Q. This is expressed more simply by an alternative formulation of equation (4.2.16);

$$Q = \sum_{\substack{\text{energy} \\ \text{levels}}} \Omega_i \exp[-\beta E_i] \qquad (4.2.18)$$

When it is convenient to employ the notation of quantum state density, the appropriate expression which properly includes one exponential term for each quantum state is:

$$Q = \int_0^\infty N(E) \exp[-\beta E]\, dE \qquad (4.2.19)$$

Combining equations (4.2.18) and (4.2.19) with (4.2.1) and (4.2.2) gives the probabilities for an assembly to be found in a state of energy E_i, or in a state with an energy between E and $E + dE$ as:

$$P(\text{energy, } E_i) = \frac{\Omega_i \exp[-\beta E_i]}{\sum\limits_{\substack{\text{energy} \\ \text{levels}}} \Omega_i \exp[-\beta E_i]} \qquad (4.2.20)$$

$$P'(\text{energy, } E)\, dE = \frac{N(E) \exp[-\beta E]\, dE}{\int_0^\infty N(E) \exp[-\beta E]\, dE} \qquad (4.2.21)$$

It should be emphasized that the actual value of Q for any specified assembly is the same however it is evaluated: the three equations (4.2.16). (4.2.18), and (4.2.19) are simply three different ways of representing the same quantity. The P's on the other hand are different for they refer to single states, energy levels and energy ranges.

We now examine the interpretation and meaning of β. If a number of assemblies are placed in contact so that they can exchange only energy and are allowed to come to equilibrium, then by the zeroth law of thermodynamics their only common property is their degree of hotness or temperature. In the molecular thermodynamic treatment we have just shown that assemblies placed in contact and able to exchange energy have a common value of a parameter β. It is therefore *reasonable* to regard β as a measure of temperature. The question arises "How is β related to our intuitive idea of temperature, and how is it related functionally to the thermodynamic temperature, T?" To answer the first part of the question we consider the internal energy of a particular sub-assembly.

The observable value of any property is the long-term average of the instantaneous value of the property, and as stated above (Section 4.1) is the properly weighted average over the appropriate ensemble. That is, the observable value of any property X, is given by

$$\bar{X} = \sum_{\text{states}} \{P(\text{state}, i) \times X_i\} \tag{4.2.22}$$

The average internal energy of an assembly, $\bar{E}$, which we identify with the internal energy of classical thermodynamics, U, is then

$$U = \bar{E} = \sum \{P(\text{state}, i) \times E_i\}$$

$$= \sum \left\{ \frac{E_i \exp\left[-\beta E_i\right]}{\sum \exp\left[-\beta E_i\right]} \right\}$$

$$= \frac{\sum E_i \exp\left[-\beta E_i\right]}{\sum \exp\left[-\beta E_i\right]} \tag{4.2.23}$$

The last expression may be simplified by examining the differential of $\ln Q$:

$$\left(\frac{\partial \ln Q}{\partial \beta}\right)_{E_i} = \frac{\partial}{\partial \beta} \{\ln\left(\exp\left[-\beta E_1\right] + \exp\left[-\beta E_2\right] + \ldots\right)\}$$

$$= -\frac{E_1 \exp\left[-\beta E_1\right] + E_2 \exp\left[-\beta E_2\right] + \ldots}{\exp\left[-\beta E_1\right] + \exp\left[-\beta E_2\right] + \ldots}$$

$$= -\frac{\sum E_i \exp\left[-\beta E_i\right]}{\sum \exp\left[-\beta E_i\right]}$$

$$= -\bar{E} = -U \tag{4.2.24}$$

In performing the differentiation it is necessary to assume that the E_i

are constant. This implies that the volume of the assembly remains constant, for under these conditions, according to quantum theory, the energies of all quantum states remain unchanged (see page 18). The internal energy is thus given by

$$U = \bar{E} = -(\partial \ln Q/\partial \beta)_{E_i \text{ or } V} \qquad (4.2.25)$$

We can now apply this formula to a particular case to obtain a more quantitative idea of the meaning of β. We note first that it is irrelevant how arbitrarily we define the model for the particular sub-assembly since the value of β is the same for all sub-assemblies in thermal equilibrium. We therefore choose a very simple model whose quantum states have energies in arithmetical progression, that is

$$E_n = nE \quad \text{where } n = 0, 1, 2, \text{ etc.} \qquad (4.2.26)$$

This formulation of energy levels is not altogether unrealistic since it is that for a simple harmonic oscillator.

The partition function then takes the form

$$Q = \exp[-0 \times \beta E] + \exp[-1 \times \beta E] + \exp[-2 \times \beta E] + \ldots$$
$$= 1 + x + x^2 + \ldots$$
$$= 1/(1 - x)$$
$$= 1/(1 - \exp[-\beta E]) \qquad (4.2.27)$$

The identification made between the second and third lines following the substitution $x = \exp[-\beta E]$, is an application of the equation for summation of a geometrical progression, or of the more general binomial theorem for the expansion of $(1 - x)^r$.

Suppose now that βE is very small compared to unity, then

$$1 - \exp[-\beta E] = 1 - \{1 - \beta E + (\beta E)^2/2! - \ldots\}$$
$$\approx \beta E \qquad (4.2.28)$$

The partition function then becomes

$$Q = 1/\beta E, \quad \ln Q = -\ln \beta - \ln E \qquad (4.2.29)$$

and internal energy is

$$\bar{E} = -(\partial \ln Q/\partial \beta)_E = +1/\beta \qquad (4.2.30)$$

The only way to change the internal energy of an assembly of fixed volume and shape is by exchange of heat with the surroundings. The application of heat to the global assembly must therefore increase the average energy of the constituent sub-assemblies, and of our special sub-assembly in particular, and its value of β must decrease. By classical thermodynamics increase of temperature of an assembly of fixed volume in equilibrium with its surroundings must be accompanied by an increase in its internal energy. Hence β must be inversely related to the accepted idea of temperature. When β is large the thermodynamic temperature must be low and vice versa. We therefore define a molecular thermodynamic scale of temperature τ such that

$$\beta = 1/k\tau \qquad (4.2.31)$$

The second part of our question, may now be stated "How is τ related to T, the thermodynamic scale of temperature?"

There are two ways in which this identification may be made. It can be carried out directly by comparing an equation derived by molecular thermodynamics with one derived from classical thermodynamics. This method makes no reference to the properties of ideal gases or other types of assembly. This method is detailed in the next few pages. Alternatively we may make the identification by developing the molecular thermodynamics of the ideal gas, or some other suitable assembly, and comparing the results with those of classical thermodynamics when applied to the same assembly. Although this method is intellectually less elegant it is simpler to appreciate, but unfortunately cannot be given until later in the book (see particularly Section 10.4).

The direct correlation of β (or τ) with T

We proceed by first obtaining the complete differential of the quantity, J, which is defined by equation (4.2.32).

$$J = k \ln Q = k \ln \left\{ \sum \exp \left[-\beta E_i \right] \right\} \qquad (4.2.32)$$

The complete differential, which allows for variations in both β and the E_i's, is

$$dJ = \left\{ \frac{\partial J}{\partial \beta} \right\}_{E_i\text{'s}} d\beta + \left\{ \frac{\partial J}{\partial E_1} \right\}_{\beta} dE_1 + \left\{ \frac{\partial J}{\partial E_2} \right\}_{\beta} dE_2 + \cdots$$

$$= \left\{ \frac{\partial J}{\partial \beta} \right\}_{E_i\text{'s}} d\beta + \sum \left\{ \frac{\partial J}{\partial E_i} \right\}_{\beta} dE_i \qquad (4.2.33)$$

Now according to the rules of wave mechanics the E_i depend only upon the volume of the assembly. We can thus write

$$dE_i = (dE_i/dV)\,dV \qquad (4.2.34)$$

whence

$$dJ = \left\{\frac{\partial J}{\partial \beta}\right\} d\beta + \sum \left\{\frac{\partial J}{\partial E_i}\right\} \left\{\frac{dE_i}{dV}\right\} dV$$

$$= k\left\{\frac{\partial \ln Q}{\partial \beta}\right\}_{E_i's} d\beta + \sum k\left\{\frac{\partial \ln Q}{\partial E_i}\right\}_{\beta} \left\{\frac{dE_i}{dV}\right\} dV \qquad (4.2.35)$$

Equation (4.2.25) identifies the first term as $-k\bar{E}\,d\beta$. What is the interpretation of the second term? The main part of second term is

$$\sum k\left\{\frac{\partial \ln Q}{\partial E_i}\right\} \left\{\frac{dE_i}{dV}\right\} = \sum \left\{k\frac{1}{Q}\frac{\partial \sum \exp[-\beta E_i]}{\partial E_i}\frac{dE_i}{dV}\right\}$$

$$= -\sum \left\{k\beta\frac{\exp[-\beta E_i]}{Q}\frac{dE_i}{dV}\right\}$$

$$= -k\beta \sum \left\{P(\text{state}, i)\frac{dE_i}{dV}\right\}$$

$$= -k\beta\,\overline{(dE/dV)} \qquad (4.2.36)$$

where the bar over (dE/dV) implies the properly weighted average of (dE_i/dV) taken over all possible quantum states for the assembly, that is over all members of the canonical ensemble. It is therefore the observable value of this quantity (see equation 4.2.22). We thus obtain

$$dJ = -k\bar{E}\,d\beta - k\beta\,\overline{(dE/dV)}\,dV \qquad (4.2.37)$$

If we replace β by $1/k\tau$, since $d\beta = -(1/k\tau^2)\,d\tau$ we obtain

$$dJ = (\bar{E}/\tau^2)\,d\tau - (1/\tau)\,\overline{(dE/dV)}\,dV \qquad (4.2.38)$$

To determine the meaning of the second term in (4.2.38) we consider the implication of the quantity dE_i/dV for a single quantum state. Suppose we have an assembly in a quantum state i whose energy is E_i, the volume of the assembly being V. By the rules of wave mechanics the energy of this quantum state can be altered only by changing the volume. When the volume is changed by an amount dV a certain amount of work is done on the assembly given by $w = -p_i\,dV$ where p_i is the pressure exerted by the assembly when in the quantum state i. By conservation of energy it follows that

$$dE_i = -p_i \, dV \quad \text{or} \quad p_i = -(dE_i/dV) \tag{4.2.39}$$

Since the observable value of any property of an assembly is the properly weighted average taken over the appropriate ensemble, the quantity $\overline{(dE/dV)}$, defined in the fourth line of equation (4.2.36), has to be identified with the observable, or average, pressure of the assembly, $\bar{p}$. We thus obtain the simple result

$$dJ = (\bar{E}/\tau^2) \, d\tau + (\bar{p}/\tau) \, dV \tag{4.2.40}$$

In classical thermodynamics an equation similar to this is obtained by differentiating $-(F/T)$ (see equations 3.2.6 and 3.2.9)

$$-d \, (F/T) = (F/T^2) \, dT - (dF/T)$$
$$= (U - TS) \, dT/T^2 + p \, dV/T + S \, dT/T$$
$$= (U/T^2) \, dT + (p/T) \, dV \tag{4.2.41}$$

In comparing equations (4.2.40) and (4.2.41) we note (a) that according to the identification made in connection with the first law of thermodynamics, $\bar{E} = U$, (b) that we have just accepted that $\bar{p} = p$ and (c) that when considering the implications of the zeroth law we showed that τ must be a monotonically increasing function of the thermodynamic temperature, T. There is therefore no escape from the identifications:

$$\tau \equiv T \quad \text{or} \quad \beta \equiv 1/kT \tag{4.2.42}$$
$$J \equiv -(F/T) + C \tag{4.2.43}$$

where C is a constant which is independent of temperature and volume. Inserting the definition of J given in equation (4.2.32), we finally obtain

$$(F/T) = -k \ln Q + C \tag{4.2.44}$$

The identifications of equations (4.2.42) and (4.2.43) correspond to the two sub-statements which are required to define the second law of thermodynamics (see Section 3.1). The entropy of the assembly is then

$$S = (U/T) - (F/T)$$
$$= kT(\partial \ln Q/\partial T)_V + k \ln Q - C \tag{4.2.45}$$

and the assembly partition function is now expressed as

$$Q = \sum_{\text{states}} \exp\left[-E_i/kT\right] \qquad (4.2.46)$$

The first term on the right-hand side of (4.2.45) arises since

$$U = \bar{E} = -(\partial \ln Q/\partial \beta)_V \qquad (4.2.47)$$

and $d\beta = -dT/kT^2$ whence $\bar{E} = kT^2 (\partial \ln Q/\partial T)_V$ \qquad (4.2.48)

The derivation to this point gives no indication of the magnitude of the constant k. Essentially k is related to the size of the unit of temperature. This is officially set by defining the temperature of water at its triple point as 273·15 K. This does not help immediately in determining k, which can in fact be found only by making quantitative measurements on some real assembly, and by comparing these results with those predicted by molecular thermodynamics. Using an ideal gas as the test assembly, it is found that in addition to the temperature datum value of 273·15 K, one requires to know the number of systems in 1 mole, that is Avogadro's number. This may be determined by experiments which are independent of any assumptions made in molecular thermodynamics. Typical of the methods for determining N_A are:

(a) measurement of the charge on the electron, and on 1 mole of univalent ions (Faraday's constant);
(b) measurement of the lattice spacing in sodium chloride using X-rays, the density of NaCl, and the formula weight;
(c) measurement of the rate of disintegration of a radioactive material giving α particles, and the rate of production of helium gas by the same material measured in moles per unit time.

The best value of k so obtained is $1·38054 \times 10^{-23}$ J K^{-1}.

The value of the constant C

Whereas k can be determined only by a specific experiment which fixes the size of the unit of temperature, the constant C is a thermodynamic quantity and is related to the zero of the scale of entropy. Its value can be assigned by using the third law of thermodynamics which for practical chemical purposes can be written

$$S_0^\circ = 0 \quad \text{(perfect crystal at 0 K)} \qquad (4.2.49)$$

Thus C is determined by evaluating the various terms in equation (4.2.45) for the limit $T = 0$. Suppose that the partition function for the assembly has the form

$$Q = \Omega_0 \exp\left[-\frac{E_0}{kT}\right] + \Omega_1 \exp\left[-\frac{E_1}{kT}\right] + \ldots$$

$$= \exp\left[-\frac{E_0}{kT}\right]\left\{\Omega_0 + \Omega_1 \exp\left[-\frac{E_1 - E_0}{kT}\right] + \ldots\right\}$$

$$= \exp\left[-\frac{E_0}{kT}\right] \times Q' \tag{4.2.50}$$

where Q' is the partition function for the assembly where the energies of the quantum states are measured from that of the lowest state rather than from some arbitrary zero of energy. Q' can be called the conventional partition function for the assembly. We then obtain for the factors which contain Q in equation (4.2.45):

$$\ln Q = -\frac{E_0}{kT} + \ln Q' \tag{4.2.51}$$

$$\frac{\partial \ln Q}{\partial T} = \frac{E_0}{kT^2} + \frac{\partial \ln Q'}{\partial T} \tag{4.2.52}$$

When T approaches zero the limiting values of $\ln Q'$ and $\partial \ln Q'/\partial T$ are

$$\underset{T \to 0}{\text{Lt}} [\ln Q'] = \underset{T \to 0}{\text{Lt}} \left\{\ln \Omega_0 + \Omega_1 \exp\left[-\frac{E_1 - E_0}{kT}\right] + \ldots\right\}$$

$$= \ln \Omega_0 \tag{4.2.53}$$

$$\underset{T \to 0}{\text{Lt}} \left[\frac{\partial \ln Q'}{\partial T}\right] = \underset{T \to 0}{\text{Lt}} \left[\frac{1}{Q'} \frac{\partial Q'}{\partial T}\right]$$

$$= \underset{T \to 0}{\text{Lt}} \frac{[0 + (1/kT^2)(E_1 - E_0)\Omega_1 \exp[-(E_1 - E_0)/kT] + \ldots]}{Q'} \tag{4.2.54}$$

In determining the value of the limit for $\partial \ln Q'/\partial T$ there is a problem for the factor $(1/kT^2)$ tends towards infinity, while the exponential factor tends to zero. The result might therefore be infinite, finite or zero at first sight. The limit is readily shown to be zero: we consider the function $(1/T^2) \exp[-A/T]$.

$$\frac{1}{T^2} \exp\left[-\frac{A}{T}\right] = \frac{1}{T^2 \exp[+A/T]}$$

$$= \frac{1}{T^2\left\{1 + \frac{A}{T} + \left(\frac{A^2}{T}\right)\frac{1}{2!} + \left(\frac{A^3}{T}\right)\frac{1}{3!} + \cdots\right\}}$$

$$= \frac{1}{T^2 + AT + \frac{A^2}{2!} + \frac{A^3}{T.3!} + \cdots} \tag{4.2.55}$$

When T tends towards zero the first two terms on the denominator become zero, but all terms after the third become infinite. The denominator therefore becomes infinite so the limit of the function is zero. Thus $\underset{T \to 0}{\text{Lt}}\, [\partial \ln Q'/\partial T] = 0$.

Inserting the values for the limits into equation (4.2.51) and (4.2.52), and then into (4.2.45) gives

$$S_0^\circ = (E_0/T + 0) + (-E_0/T + k \ln \Omega_0) - C$$

$$= k \ln \Omega_0 - C \tag{4.2.56}$$

By the third law the entropy of a perfect crystal at 0 K, S_0°, is zero. Therefore

$$C = k \ln \Omega_0 \quad \text{(perfect crystal)} \tag{4.2.57}$$

We must now decide about the degeneracy of the lowest quantum state for a perfect crystal at 0 K. A perfect crystal is one in which all the systems are identical, and in which there are no defects in the crystal structure. In quantum mechanical terms it would appear that at 0 K every system would have to be in its lowest quantum state with say a wave function ψ_0. The assembly wave function would then be

$$\Psi_0 = \psi_0(a)\psi_0(b)\psi_0(c)\ldots\psi_0(n) \tag{4.2.58}$$

Since all the ψ_0's would be identical functions of the position of the systems relative to say the centre of the unit cell, there would be no possible permutations of the ψ_0's amongst the various sites, $a, b, c, \ldots n$ of the crystal which could produce distinguishable new quantum states of the assembly. Thus the ground state of the crystal must be non-degenerate, that is

$$\Omega_0 = 1 \quad \text{whence} \quad C = k \ln \Omega_0 = 0 \qquad (4.2.59)$$

The third law thus enables us to write a molecular thermodynamic equation which gives the correct analogue of the classical thermodynamic entropy, namely

$$S = kT(\partial \ln Q/\partial T)_V + k \ln Q$$
$$= (\partial/\partial T)(kT \ln Q)_V \qquad (4.2.60)$$

The free energy is then

$$F = -kT \ln Q \qquad (4.2.61)$$

This equation appears at first sight to give an absolute value for the free energy F. Actually this is not so since Q will always contain the arbitrary factor $\exp[-E_0/kT]$, and so F contains an arbitrary term E_0. Just as there is no way of fixing an absolute zero of internal energy, there is no way of fixing an absolute zero of free energy.

We conclude this section by returning to the pressure of an assembly. According to classical thermodynamics

$$p = -(\partial F/\partial V)_T \qquad (4.2.62)$$

Inserting equation (4.2.61) for F then gives

$$p = kT(\partial \ln Q/\partial V)_T \qquad (4.2.63)$$

Differentiating equation (4.2.47), at constant temperature, and remembering that the E_i are functions of volume we obtain

$$p = \frac{kT}{Q} \frac{\partial}{\partial V} \left\{ \sum \exp\left[-\frac{E_i}{kT}\right] \right\}$$

$$= -\frac{kT}{Q} \sum \left\{ \frac{1}{kT} \frac{dE_i}{dV} \exp\left[-\frac{E_i}{kT}\right] \right\}$$

$$= -\sum \frac{dE_i}{dV} P(\text{state}, i)$$

$$= \overline{(dE/dV)} \qquad (4.2.64)$$

The final result, as it must be, is the same as previously used in the identification of τ and T.

4.3 THE CLASSICAL PARTITION FUNCTION

The theory of the classical assembly in a thermostat may be developed by arguments similar to those employed for the conceptually simpler quantum assembly. Whereas for an isolated assembly only those regions of phase space corresponding to a total energy E were accessible, for an assembly in a thermostat all regions become accessible.

If we denote by $p_1, p_2, \ldots, p_M$ and $q_1, q_2, \ldots, q_M$ the momenta and positions of the systems of the assembly, then a typical element in phase space is obtained by specifying infinitesimal ranges for these coordinates, that is p_1 to $p_1 + \mathrm{d}p_1$, p_2 to $p_2 + \mathrm{d}p_2$ and so on. The volume of such an element in phase space is

$$\mathrm{d}V = \mathrm{d}p_1 \, \mathrm{d}q_1 \, \mathrm{d}p_2 \, \mathrm{d}q_2 \ldots \mathrm{d}p_M \, \mathrm{d}q_M \qquad (4.3.1)$$

In this expression M is the total number of positional coordinates required to specify the precise configuration of all the systems of the assembly. Thus if each system contained n atoms, then for an assembly of N systems $M = 3.n.N$.

The energy associated with any position in phase space is a function of the p's and q's (for an ideal gas of structureless systems it is a function only of the p's). Since equal probabilities attach to equal volumes of phase space of the same energy, we can write after developing arguments similar to those in Section 4.2,

$$P(\text{state}, \mathrm{d}V) = \frac{\exp\left[-E/kT\right] \mathrm{d}p_1 \, \mathrm{d}q_1 \, \mathrm{d}p_2 \, \mathrm{d}q_2 \ldots \mathrm{d}p_M \, \mathrm{d}q_M}{\int \ldots \int\limits_{p,q} \exp\left[-E/kT\right] \mathrm{d}p_1 \, \mathrm{d}q_1 \ldots \mathrm{d}p_M \, \mathrm{d}q_M}$$

$$(4.3.2)$$

The denominator of this expression is the classical partition function, that is

$$Q_{\text{classical}} = \int \ldots \int \exp\left[-E/kT\right] \mathrm{d}p_1 \, \mathrm{d}q_1 \ldots \mathrm{d}p_M \, \mathrm{d}q_M \quad (4.3.3)$$

where the multiple integral is evaluated for the whole accessible region of phase space.

We saw earlier that the classical formulation in terms of phase space could be related to the quantum formulation if the phase space was suitably divided into cells of volume h^M when the systems of the assembly were localized, and $h^M \times N!$ if the systems were non-localized. Each cell then corresponds to a single quantum state. The value of the classical partition function can thus be made to correspond with that of the quantum partition function by dividing by the relevant factor. Thus

$$Q_{\substack{\text{semi-classical} \\ \text{(crystal)}}} = h^{-M} \int \dots \int \exp \left[-\frac{E}{kT} \right] \mathrm{d}V \qquad (4.3.4)$$

$$Q_{\substack{\text{semi-classical} \\ \text{(gas)}}} = \frac{h^{-M}}{N!} \int \dots \int \exp \left[-\frac{E}{kT} \right] \mathrm{d}V \qquad (4.3.5)$$

where the symbol $\mathrm{d}V$ is defined by equation (4.3.1).

When quantum effects are not important equations (4.3.4) and (4.3.5) give the same numerical value of Q as the quantum formulation. If quantum effects are important they give larger values.

The use of equations (4.3.4) and (4.3.5) does not of course affect the probability that an assembly is found in a volume $\mathrm{d}V$ of phase space for both numerator and denominator of equation (4.3.2) are then divided by the same factor.

4.4 SUMMARY OF IMPORTANT EQUATIONS

In this chapter we have derived equations which relate the thermodynamic properties of an assembly in a thermostat to the energies of the quantum states allowed to the assembly. These equations are basic to molecular thermodynamics and are as follows:

The assembly partition function Q

—quantum form

$$Q = \sum_{\text{states}} \exp \left[-E_i/kT \right] \qquad (4.4.1)$$

$$Q = \sum_{\substack{\text{energy} \\ \text{levels}}} \Omega_i \exp \left[-E_i/kT \right] \qquad (4.4.2)$$

$$Q = \int_0^\infty N(E) \exp \left[-E/kT \right] \mathrm{d}E \qquad (4.4.3)$$

—classical form,

$$Q = \int \dots \int_{p,q} \exp \left[-E/kT \right] \mathrm{d}p_1 \, \mathrm{d}p_2 \dots \mathrm{d}p_M \, \mathrm{d}q_M \qquad (4.4.4)$$

Internal energy, $U = \bar{E}$,

$$U = \bar{E} = kT^2 \left(\partial \ln Q/\partial T \right)_{E_i \text{ or } V} \qquad (4.4.5)$$

Free energy, F,

$$F = -kT \ln Q \qquad (4.4.6)$$

Entropy, S,

$$S = kT(\partial \ln Q/\partial T)_{E_i \text{ or } V} + k \ln Q$$

$$= \left\{ \frac{\partial(kT \ln Q)}{\partial T} \right\}_{E_i \text{ or } V} \qquad (4.4.7)$$

PROBLEMS

4.1 Consider the assembly of example 1.2. Are any of the states of energy 6ϵ more probable than the others? Are any of the possible energy distributions with a total energy 6ϵ more probable than the others?

4.2 Referring to the first paragraph of Section 4.2, consider qualitatively the magnitude of energy fluctuations from the mean as the assembly is reduced in size until it eventually becomes a single molecule.

4.3 Outline the argument in Section 4.2 from equation (4.2.1) to (4.2.17) in terms of weakly coupled localized systems instead of assemblies. Thus derive the concept of, and equation for, the molecular partition function.

4.4 The following figures give the height distribution of 450 men taken at random. Tabulate the probabilities for each height range, and use equation (4.2.22) to obtain the mean height of the group. Check by evaluating the mean directly, that is by computing the total height of all men and dividing by 450.

Height/ft	4·9	5·1	5·3	5·5	5·7	5·9	6·1	6·3	6·5	6·7
Number	1	5	21	58	105	131	94	30	4	1

4.5 Using the equation $\bar{E} = kT^2(\partial \ln Q/\partial T)_V$ obtain an expression for the heat capacity at constant volume, $C_V = (\partial \bar{E}/\partial T)_V$ in terms of Q.

4.6 Energy fluctuations. The best measure of the magnitude of energy fluctuations in a thermostatted assembly is the root-mean-square deviation from the long term mean, that is

$$\sigma_E = \{\overline{\delta E^2}\}^{1/2} \equiv \{\overline{(E_i - \bar{E})^2}\}^{1/2} \qquad \text{(i)}$$

where E_i is the energy of any state of the assembly, and the average is taken over the canonical ensemble.

Prove first that equation (i) is equivalent to

$$\sigma_E^2 = \overline{E^2} - (\bar{E})^2 \qquad \text{(ii)}$$

Hence obtain σ_E^2 by differentiating both sides of the identity (iii) with respect to β

$$\bar{E} \times Q \equiv \sum_{\text{states}} E_i \exp\left[-\beta E_i\right] \qquad \text{(iii)}$$

Thereby prove that

$$\sigma_E/\bar{E} = \text{fractional energy fluctuation}$$

$$= \frac{(kC_v)^{1/2}}{\bar{E}/T} \qquad \text{(iv)}$$

4.7   For an ideal monatomic gas $\bar{E}/T = \frac{3}{2}Nk$, where N is the number of atoms in the gas. Using the result of problem 4.6, obtain a simple expression for $\sigma_E/\bar{E}$. Show that this is always negligible for assemblies of macroscopic size. What volume of Ar at S.T.P. would show energy fluctuations of about 1 in 10^6?

CHAPTER 5

MOLECULAR AND ASSEMBLY PARTITION FUNCTIONS

Equations (4.4.5) to (4.4.7) give the major thermodynamic functions in terms of the assembly partition function, Q. If the assembly partition function can be expressed in terms of molecular parameters, the aim of molecular thermodynamics can be achieved, namely the formulation of thermodynamic properties in terms of the properties of the molecules themselves. This can be done simply only if the systems of the assembly are independent and have well defined molecular quantum states, for then and only then can we define a molecular partition function which can later be related to the assembly partition function. The molecular partition function can be arrived at by the following reasoning.

When dealing with the assembly in a thermostat we made no assumption about what was meant by an assembly, and we could, for example, have made one of the sub-assemblies within the global assembly a single molecule. We should then have arrived at the result that the probability that a molecule was to be found in a quantum state of energy ϵ_j was

$$p(\text{state}, \epsilon_j) = \frac{\exp\left[-\epsilon_j/kT\right]}{\sum\limits_{\text{states}} \exp\left[-\epsilon_i/kT\right]} \tag{5.0.1}$$

Proceeding as before we should have defined the molecular partition function as

$$q = \sum_{\text{states}} \exp\left[-\frac{\epsilon_i}{kT}\right] \tag{5.0.2}$$

$$= \sum_{\substack{\text{energy} \\ \text{levels}}} g_i \exp\left[-\frac{\epsilon_i}{kT}\right]$$

where g_i is the degeneracy of the ith molecular energy level.

In the context of an assembly of systems, the molecular partition function has meaning only when all the systems of a given type possess the same set of quantum states, and when the energy of a system in any particular quantum state is unaffected by the energies of the remaining systems of the assembly. This is true only when the systems are independent. Although this is never strictly true, two types of real assembly approximate to this ideal, namely highly dilute gases and crystals of monatomic solids. Of these two the first approximates well, while the latter rather poorly, The usefulness of the molecular partition function is thus limited to two types of ideal assembly, assemblies of independent localized systems and assemblies of independent non-localized systems. When it becomes necessary to consider non-ideal gases and crystals, the assembly partition must be used, and the simple relationships which we now derive in Sections 5.1 and 5.2 are no longer applicable. For the treatment of such assemblies the reader is referred to more advanced texts.

5.1 THE ASSEMBLY OF INDEPENDENT LOCALIZED SYSTEMS: THE IDEAL CRYSTAL

As shown in Chapter 2, the quantum state of an assembly of localized systems is defined by assigning molecular or system wave functions to each site in the assembly. Thus the three states of the 12 system assembly shown in Figure 5.1 have the same energy but represent distinct quantum states or complexions of the assembly. In constructing the assembly

(a)   (b)   (c)

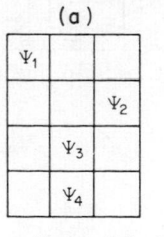 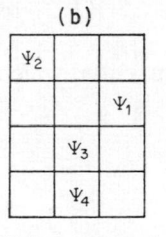 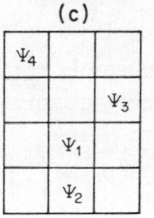

Figure 5.1 Three different complexions or quantum states of an assembly of 12 localized systems. The ψ's of the unmarked sites are the same for (a), (b) and (c).

partition function it is necessary to include one term of the form $\exp\left[-E/kT\right]$ for each distinct quantum state or complexion.

To derive the relationship between Q and q we first consider a rudimentary assembly consisting of two systems at two sites a and b. We further limit the assembly by allowing only three quantum states to be accessible to the systems at each site. Thus the allowed molecular wave functions and energies can be represented as ψ_1, ψ_2, ψ_3 and ϵ_1, ϵ_2, ϵ_3. There are then 9 possible assignments of quantum states to the two-system assembly, or 9 distinct complexions of the assembly.

Assembly state number	Assignment of ψ's to sites (a)	(b)	Assembly wave function	Energy of assembly
1	ψ_1	ψ_1	$\psi_1(a)\psi_1(b)$	$2\epsilon_1 = E_1$
2	ψ_1	ψ_2	$\psi_1(a)\psi_2(b)$	$\epsilon_1 + \epsilon_2 = E_2$
3	ψ_1	ψ_3	$\psi_1(a)\psi_3(b)$	$\epsilon_1 + \epsilon_3 = E_3$
4	ψ_2	ψ_1	$\psi_2(a)\psi_1(b)$	$\epsilon_1 + \epsilon_2 = E_4$
5	ψ_2	ψ_2	$\psi_2(a)\psi_2(b)$	$2\epsilon_2 = E_5$
6	ψ_2	ψ_3	$\psi_2(a)\psi_3(b)$	$\epsilon_2 + \epsilon_3 = E_6$
7	ψ_3	ψ_1	$\psi_3(a)\psi_1(b)$	$\epsilon_3 + \epsilon_1 = E_7$
8	ψ_3	ψ_2	$\psi_3(a)\psi_2(b)$	$\epsilon_3 + \epsilon_2 = E_8$
9	ψ_3	ψ_3	$\psi_3(a)\psi_3(b)$	$2\epsilon_3 = E_9$

For this assembly the energy levels $2\epsilon_1$, $2\epsilon_2$ and $2\epsilon_3$ are non-degenerate while the levels $\epsilon_1 + \epsilon_2$, $\epsilon_2 + \epsilon_3$ and $\epsilon_3 + \epsilon_1$ are doubly degenerate.

The molecular partition function is

$$q = \exp\left[-\frac{\epsilon_1}{kT}\right] + \exp\left[-\frac{\epsilon_2}{kT}\right] + \exp\left[-\frac{\epsilon_3}{kT}\right] \tag{5.1.1}$$

and the assembly partition function, which must contain one term for each assembly quantum state, is

$$Q = \sum_{\substack{\text{states} \\ 1-9}} \exp\left[-\frac{E_i}{kT}\right]$$

$$= \exp\left[-\frac{2\epsilon_1}{kT}\right] + \exp\left[-\frac{2\epsilon_2}{kT}\right] + \exp\left[-\frac{2\epsilon_3}{kT}\right] + 2\exp\left[-\frac{\epsilon_1 + \epsilon_2}{kT}\right]$$

$$+ 2\exp\left[-\frac{\epsilon_2 + \epsilon_3}{kT}\right] + 2\exp\left[-\frac{\epsilon_3 + \epsilon_1}{kT}\right]$$

$$= \left(\exp\left[-\frac{\epsilon_1}{kT}\right] + \exp\left[-\frac{\epsilon_2}{kT}\right] + \exp\left[-\frac{\epsilon_3}{kT}\right]\right)^2$$

$$= q^2 \tag{5.1.2}$$

The identification made between the second and third lines of equation (5.1.2) is easily verified by expanding the square. It is readily seen that increasing the number of accessible quantum states has no influence on the final result, $Q = q^2$. However if the number of systems were increased to three the assembly partition function would be of the general form

$$\sum_i \exp\left[-\frac{E_i}{kT}\right] = \sum_{k, l, m} \exp\left[-\frac{\epsilon_k + \epsilon_l + \epsilon_m}{kT}\right] \qquad (5.1.3)$$

It is not difficult to see that q^3 will produce a sum which contains one term for every possible complexion of a three-system assembly. We thus arrive at the general result for N systems

$$Q = q^N, \qquad \ln Q = N \ln q \qquad (5.1.4)$$

The thermodynamic functions for an assembly of N localized systems thus take the forms (see equations 4.4.5 to 4.4.7):

$$U = kT^2(\partial \ln Q/\partial T)_V$$

$$= NkT^2(\partial \ln q/\partial T)_V \qquad (5.1.5)$$

$$F = -NkT \ln q \qquad (5.1.6)$$

$$S = NkT(\partial \ln q/\partial T)_V - Nk \ln q \qquad (5.1.7)$$

The molar heat capacity at constant volume is

$$C_v = (\partial U/\partial T)_V$$

$$= 2NkT(\partial \ln q/\partial T)_V + NkT^2(\partial^2 \ln q/\partial T^2)_V \qquad (5.1.8)$$

5.2   THE ASSEMBLY OF INDEPENDENT NON-LOCALIZED SYSTEMS: THE IDEAL GAS

An important property of all fluids which are commonly regarded as gases is that the number of quantum states available to the molecules greatly exceeds the number of molecules in the gas. To be more precise the number of states with energies below the typical thermal energy kT greatly exceeds the number of molecules in the gas. Thus the molecular quantum states for translation are very sparsely occupied in any normal gas.

This property of gases is so important for the derivation now to be given that we justify it before proceeding further. If the translational states are not sparsely occupied, the molecular thermodynamic treatment of the gas is greatly complicated.

The allowed energy levels for translation in a cube of side L are given by equation (5.2.1) (see equation 2.1.8)

$$\epsilon = (h^2/8mL^2)(j^2 + l^2 + n^2) \tag{5.2.1}$$

where j, l and n can be any integers. Suppose that we represent the quantum numbers by distances along the x-, y-, and z-axes, then any particular quantum state is represented by a point (j, l, n) with integral coordinates in this "quantum number space". To find the number of states whose energies are below a specified energy ϵ_0 we have to count the number of points with integral coordinates within a surface corresponding to ϵ_0. Since each point is associated with unit volume in quantum number space, the number of such points is equal to the volume enclosed by the surface corresponding to the energy ϵ_0.

Now equation (5.2.1) may be rewritten

$$j^2 + l^2 + n^2 = (8m\epsilon)(L^2/h^2) \tag{5.2.2}$$

which is seen to represent a sphere of radius $R = (8m\epsilon)^{1/2}(L/h)$. The number of quantum states with energies less than ϵ_0 is then the volume of the positive octant of the sphere of radius R_0, the positive octant because j, l and n can have only positive values. The situation is illustrated for two dimensions in Figure 5.2, where the count is over the positive quadrant of a circle of radius $R_0 = 10$. The number of quantum states with energy below kT for translation in three dimensions is then

$$Z = (\tfrac{1}{8})(\tfrac{4}{3})\pi R_0{}^3 = \pi R_0{}^3/6 \tag{5.2.3}$$

Substituting $R_0 = (8m\epsilon_0)^{1/2}(L/h)$ and setting $\epsilon_0 = kT$ gives

$$Z = \frac{\pi}{6}\left\{\frac{8mkT}{h^2}\right\}^{3/2} L^3 = \frac{4}{3\sqrt{\pi}}\left\{\frac{2\pi mkT}{h^2}\right\}^{3/2} V \tag{5.2.4}$$

where V is the volume of the cube. If MW = molecular weight of the gas then the weight of a single molecule, m in kg, is

$$m = MW/L_A \tag{5.2.5}$$

where $L_A = 6{\cdot}02 \times 10^{26}$ kmol^{-1}

Inserting (5.2.5) into (5.2.4) gives

$$Z = \frac{4}{3\sqrt{\pi}}\left\{\frac{2\pi MWkT}{L_A h^2}\right\}^{3/2} V \tag{5.2.6}$$

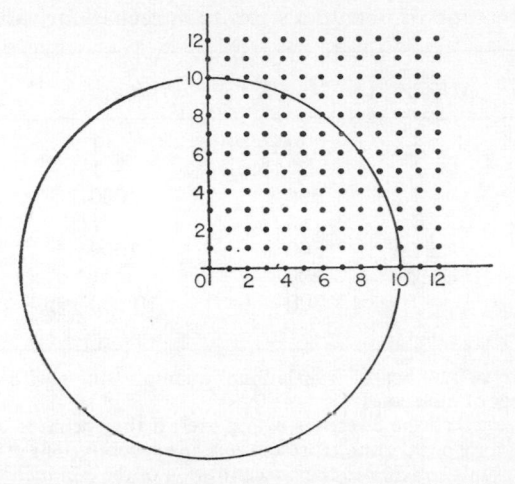

Figure 5.2 Quantum number space for two-dimensional translation. Seventy-eight quantum states occur within the positive quadrant of the circle and have energies below $100h^2/8mL^2$. The area of the quadrant is 78·5 units, $R_0 = 10$.

The number of molecules in a volume, V, according to the ideal gas law is

$$N = \frac{L_A 273 PV}{22 \cdot 4T} \qquad (5.2.7)$$

where P is the pressure in atmospheres, V is in m^3, and T in K. The ratio of states to molecules is then

$$\frac{Z}{N} = \left\{ \frac{4\pi(2k)^{3/2} \times 22 \cdot 4}{3 \times h^3 \times 273 \times L_A^{3/2}} \right\} T^{5/2} M W^{3/2} P^{-1}$$

$$= (2 \cdot 0 \times 10^{-2}) \times T^{5/2} M W^{3/2} P^{-1} \qquad (5.2.8)$$

Table 5·1 shows values of Z/N for a number of gases at various temperatures and pressures. For all substances normally regarded as gases the ratio considerably exceeds unity. Even for helium at its boiling point $Z/N = 5$; for all other gases it is much larger being typically about 10^5 at S.T.P.

An important type of assembly for which Z/N may be well below unity is "electron gas". Because of its exceedingly low mass, the number of translational states for an electron with energy below kT

Table 5.1 The ratio of quantum states to molecules for various gases

Molecules	MW/amu	T/K	P/atm	Z/N^a
He	4	4 (b.p.)	1	5
H_2	2	27 (b.p.)	1	130
H_2	2	300	1000	90
N_2	28	80 (b.p.)	1	$1 \cdot 7 \times 10^5$
N_2	28	300	1000	$4 \cdot 5 \times 10^3$
electron	1/1850	300	1^b	$0 \cdot 4$
		300	$10\ 000^{b,\,c}$	4×10^{-5}

^{a} Z/N = ratio of number of translational quantum states with energies below kT to the number of molecules.

^{b} The pressure which the electrons would exert if they behaved as an ideal gas. These pressures may be taken to represent certain concentrations of electrons.

^{c} This corresponds roughly to the concentration of the conductivity electrons in a metal.

is comparable to the number of electrons when their concentration corresponds to an ideal gas pressure of 1 atm. While one does not normally consider electrons to exist as a gas at this sort of concentration, the conductivity electrons in metals, being free to move throughout the metal, behave in many respects as a gas and their concentration would correspond to an ideal gas pressure of the order of 10^4 atm. Under these conditions the number of electrons greatly exceeds the number of low-lying quantum states. Since only two electrons may occupy the same translational quantum state, the pair having opposed spins, a situation arises in which the translational quantum states are occupied to energies greatly exceeding kT. There is then a sharp transition from completely filled quantum states to virtually empty quantum states over an energy range of a few kT. This is illustrated diagrammatically in Figure 5.3. This feature of electron gas accounts for several important properties of electrons in metals, such as the absence of any electronic heat capacity.

Another type of assembly where the number of systems may well exceed the number of low lying quantum states is photon gas, or blackbody radiation, but here there is no restriction on the number of systems which can occupy the same quantum state. The distribution of photons amongst quantum states with energies around kT is thus very different from that of electrons, as seen from Figure 5.3.

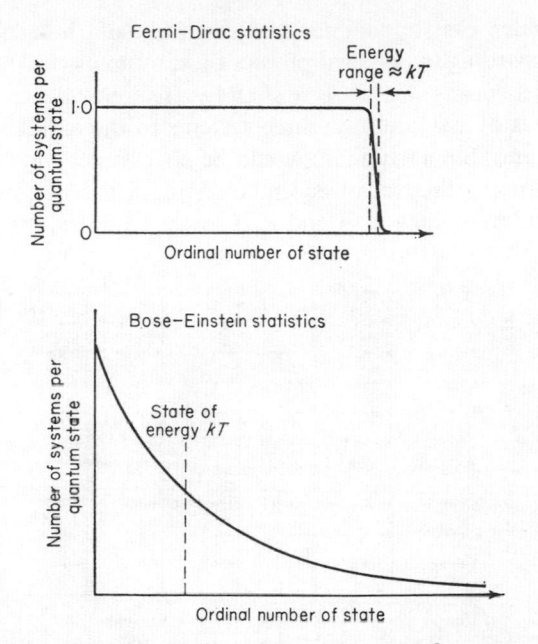

Figure 5.3 Occupation of quantum states for systems obeying Fermi–Dirac statistics and Bose–Einstein statistics.

All species giving assembly wave functions which are antisymmetric with respect to exchange of systems (that is those containing odd numbers of fundamental particles) behave like electrons, and are said to obey Fermi–Dirac statistics: they are called Fermions. All species giving assembly wave functions which are symmetric with respect to exchange of systems (that is those containing even numbers of fundamental particles) behave like photons, and are said to obey Bose Einstein statistics: they are called Bosons. The development of these forms of statistics is beyond the scope of this book, but is covered in advanced monographs (but see Section 5.4).

When the translational quantum states are sparsely occupied, as they are for most gases, Fermi–Dirac and Bose–Einstein statistics converge and become indistinguishable from the much simpler Maxwell–Boltzmann statistics for which we now develop the relationship between the assembly partition function, Q, and the molecular partition function, q.

We consider first a rudimentary assembly of three non-localized systems. According to classical physics these systems could, in principle be distinguished, say, by labels α, β and γ. There would then be $3! = 6$ possible ways of assigning the three systems to three different specified quantum states. Similarly there would be six distinct ways of assigning quantum states to three localized (but indistinguishable) systems occupying three different sites a, b and c. The six complexions for the two

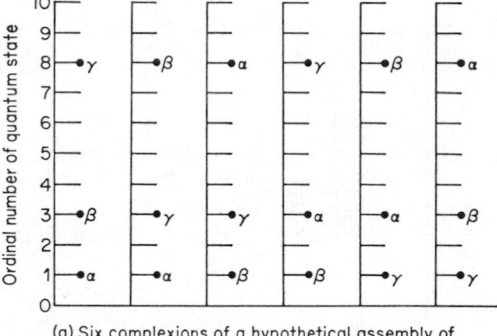

(a) Six complexions of a hypothetical assembly of three distinguishable non-localized systems occupying quantum states 1, 3 and 8

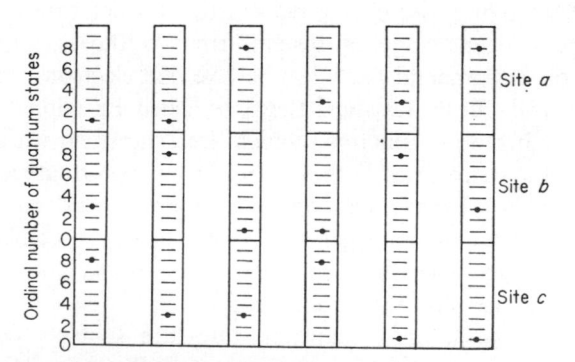

(b) Six complexions of an assembly of three indistinguishable localized systems occupying molecular quantum states 1, 3 and 8

Figure 5.4 Complexions of assemblies of distinguishable non-localized systems and of indistinguishable localized systems.

types of assembly are illustrated in Figure 5.4. Any energy level of either assembly, which is defined by stating which system quantum states are occupied, is six-fold degenerate. Quantum mechanics insists, however, that systems of any specified sort are absolutely indistinguishable, and therefore that the six supposed complexions of the assembly of non-localized systems shown in Figure 5.4(a) really correspond to a unique quantum state of the assembly which can be completely defined by saying only that quantum states 1, 3 and 8 are occupied.

Extending the argument to assemblies of N systems occupying N different specified quantum states, we conclude that there will be $N!$ complexions for assemblies containing *either* N distinguishable systems within a common enclosure, *or* N indistinguishable systems at distinguishable sites. We can use the proof given in Section 5.1 to show that the molecular and assembly partition functions are related in the same way for both types of assembly, namely

$$Q = q^N \quad \text{(assembly of indistinguishable localized systems at distinguishable sites)} \tag{5.2.9}$$

$$Q = q^N \quad \text{(assembly of distinguishable non-localized systems in a common enclosure)} \tag{5.2.10}$$

Thus any energy level for either type of assembly, when all occupied quantum states are different, will be $N!$-fold degenerate. However, since the systems of any assembly, whether localized or not, are absolutely indistinguishable, (5.2.10) is incorrect for any real assembly of non-localized systems. It incorporates the error that all energy levels of the assembly are taken to be $N!$ degenerate, when, in fact, they are non-degenerate. The assembly partition function as given by (5.2.10) is therefore overestimated by a factor of $N!$, and the correct expression is

$$Q = q^N/N! \quad \text{(assembly of indistinguishable systems in a common enclosure)} \tag{5.2.11}$$

The condition that the systems of the assembly occupy different molecular quantum states is seen to be necessary to produce the factor $N!$ As we have shown this is true for almost all substances which we should normally class as gases. It is only for helium near its boiling point that it may be an unwarranted oversimplification.

When N is large, as it is for all practical assemblies, $N!$ may be approximated by "Stirling's approximation", the most accurate form of which is:

$$\ln N! = N \ln N - N + \tfrac{1}{2} \ln (2\pi N) \qquad (5.2.12)$$

A less accurate approximation is readily derived as follows. By definition

$$\ln N! = \ln 1 + \ln 2 + \ln 3 + \ldots + \ln (N - 1) + \ln N \qquad (5.2.13)$$

This summation is represented by the area under the stepped curve in Figure 5.5 taken up to $r = (N + 1)$. The smooth curves to the left and right of the stepped curve have the equations

$$y = \ln r \qquad (5.2.14)$$

$$y = \ln (r - 1) \qquad (5.2.15)$$

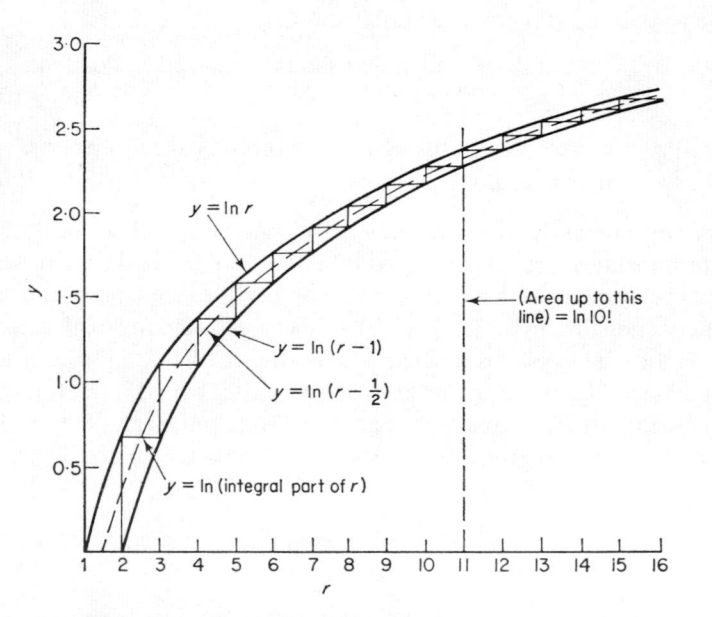

Figure 5.5 Approximations for $\ln N!$ Area under stepped curve $= \ln N!$; area under broken curve $= 1 + (N + \tfrac{1}{2}) \ln (N + \tfrac{1}{2}) - (N + \tfrac{1}{2})$. See also Table 5.2.

and clearly have areas greater than, and less than that of the stepped curve. The true area under the stepped curve must be somewhere between the extremes and will be much better approximated by the area under the broken curve defined by equation (5.2.16)

$$y = \ln\{r - \tfrac{1}{2}\} \tag{5.2.16}$$

We then obtain

$$\ln N! \approx \int_{3/2}^{N+1} \ln(r - \tfrac{1}{2})\,\mathrm{d}r$$

$$= \int_{1}^{N+1/2} \ln x\,\mathrm{d}x$$

$$= [x \ln x - x]_1^{N+1/2}$$

$$= \{N + \tfrac{1}{2}\}\ln\{N + \tfrac{1}{2}\} - \{N + \tfrac{1}{2}\} + 1 \tag{5.2.17}$$

For the majority of formulae in molecular thermodynamics we require not Q itself but $\ln Q$, and it is the accuracy of $\ln N!$ which is important, not that of $N!$. When N is very large, the difference between the approximation (5.2.17) and the less precise approximation (5.2.18) becomes unimportant

$$\ln N! \approx N \ln N - N \tag{5.2.18}$$

The three approximations are compared with the exact value of $\ln N!$ in Table 5.2. Equation (5.2.12) gives $N!$ to within 0.1% when $N = 100$

Table 5.2 Approximations for $\ln N!$

Equation		(5.2.18)	(5.2.17)	(5.2.12)
N	$\ln N!$ (exact)	$N \ln N - N$	$(N+\tfrac{1}{2})\ln(N+\tfrac{1}{2})$ $-N + \tfrac{1}{2}$	$N \ln N - N$ $+\tfrac{1}{2}\ln(2\pi N)$
3	1·791	0·296	1·885	1·764
6	6·579	4·750	6·666	6·565
10	15·104	13·026	15·189	15·096
20	42·336	39·915	42·419	42·331
40	110·321	107·555	110·403	110·318
100	363·739	360·517	363·821	363·738
200		859·663	863·313	863·232
500		2607·304	2611·412	2611·330
1000		5907·755	5912·209	5912·128
2000		13 201·80	13 206·61	13 206·52
5000		37 586·97	37 591·22	37 591·14
10 000		82 102·40	82 109·01	82 108·92

whereas (5.2.18) gives $N!$ low by a factor of about $\sqrt{(2\pi N)}$. Although this is a large number when N is of the order of 10^{20}, the proportional error in $\ln N!$ is negligible. Equation (5.2.18) is therefore completely adequate in those applications of molecular thermodynamics where $\ln Q$ is the relevant parameter.

We thus obtain

$$\ln Q = N \ln q - N \ln N + N \quad \text{(assembly of localized systems)}$$
$$(5.2.19)$$

The key thermodynamic functions may then be expressed using equations (4.4.5) to (4.4.7) as

$$U = \bar{E} = NkT^2 \left(\partial \ln q / \partial T\right)_V \tag{5.2.20}$$

$$F = -NkT \ln q + NkT \ln N - NkT \tag{5.2.21}$$

$$S = NkT \left(\partial \ln q / \partial T\right)_V + Nk \ln q - Nk \ln N + Nk \tag{5.2.22}$$

Comparison of equations (5.2.22) and (5.1.7) might suggest that, contrary to experience, the entropy of a solid should be much greater than that of a gas because of the effect of subtracting the terms ($Nk \ln N - Nk$) which will certainly be large for any practical assembly. This suggestion is in fact based upon a misconception. Although ($Nk \ln N - Nk$) is large, the molecular partition function for a gas is vastly greater than for a solid. The difference arises largely because of the small volume accessible to the motion of an atom in a solid compared to that accessible for motion of a molecule in a gas. The energy levels of the solid are widely spaced and q is small, while those for a gas are very numerous and closely spaced, which makes q large. Typically q for a solid might be of the order of unity, while q for a gas might be 10^{32} for a volume of 1 m^3 (see Section 6.1). In practice the difference between $Nk \ln q_{gas}$ and $Nk \ln q_{solid}$ turns out to be more than adequate to compensate for terms like ($Nk \ln N - Nk$) which appear in the equations for the entropy and free energy of a gas.

5.3 MULTIPLICATION THEOREM FOR PARTITION FUNCTIONS

For many types of system the total energy may be regarded to a close approximation as the sum of independent contributions from different modes of motion. Thus we may often write

$$\epsilon_{total} = \epsilon_{trans} + \epsilon_{rot} + \epsilon_{vib} \tag{5.3.1}$$

This equation is accurate if there is no coupling between the different modes of motion. Because of the laws of conservation of momentum, angular momentum and energy, there cannot be coupling between the translational motion of a body and its internal motions when moving in a field free environment. However, by the same laws there is inevitably some coupling between rotations and vibrations and there is generally coupling between different vibrations.

Consider, for example, the nitrogen molecule. The moment of inertia of the molecule is μR^2, where μ is the reduced mass, and R the bond length (internuclear separation). The kinetic energy of rotation is then

$$\epsilon_{rot} = I\omega^2/2 = \mu R^2 \omega^2/2 \qquad (5.3.2)$$

where ω is the angular velocity. As the molecule vibrates, R oscillates about its mean value, the amplitude being greater the greater the energy of the vibration. Since by the law of conservation of angular momentum, $I\omega$ must remain constant, both I and ω must change as the molecule vibrates, and the energy of rotation $I\omega^2/2$ will change in phase with the vibration. Similarly any change in the angular momentum will alter the centrifugal force and so perturb the vibration of the molecule by causing what appears as a change in the force constant with displacement from equilibrium. The interaction of rotation and vibration thus causes the vibration to become anharmonic, and since the effective force constant decreases with the vigour of rotation the frequency of oscillation will be lower in a rapidly rotating molecule than in a slowly rotating molecule.

Fortunately coupling effects of this sort are generally slight and in an elementary treatment may be ignored.

Coupling between different vibrational modes is nearly always present in real polyatomic molecules, since no vibrational modes are perfectly harmonic. The potential energy generally depends upon displacement according to a relation approximating to the Morse equation (see Section 6.3). This is equivalent to saying that the force constant in the equation of motion is dependent upon displacement. According to classical mechanics there can be a cyclic interchange of energy between anharmonic oscillators, the rate of interchange of energy being proportional to the degree of coupling which is related to the anharmonicity. But once again the consequences of these effects are generally slight as far as equilibrium thermodynamic properties are concerned, and we shall ignore them.

When the modes of motion of a molecule may be assumed independent the wave function for the molecule as a whole can be factorized into the wave functions for the independent modes of motion. This arises since the potential function in the Schrödinger equation can be expressed as a sum of independent terms. Thus

$$\psi_{\text{total}} = \psi_{\text{trans}} \, \psi_{\text{rot}} \, \psi_{\text{vib}} \qquad (5.3.3)$$

Formally the situation is not very different from that in an assembly of independent localized systems (see equation 2.1.19). The molecular partition function can then be written

$$q = \sum_i \sum_j \sum_k \exp\left[\frac{\epsilon_{i(\text{trans})} + \epsilon_{j(\text{rot})} + \epsilon_{k(\text{vib})}}{kT}\right]$$

$$= \sum_i \exp\left[-\frac{\epsilon_{i(\text{trans})}}{kT}\right] \times \sum_j \exp\left[-\frac{\epsilon_{j(\text{rot})}}{kT}\right]$$

$$\times \sum_k \exp\left[-\frac{\epsilon_{k(\text{vib})}}{kT}\right]$$

$$= q_{\text{trans}} q_{\text{rot}} q_{\text{vib}} \qquad (5.3.4)$$

The second line follows from the first since the product of the typical terms in the second line produces the typical term in the first line, and since the first line allows for all possible combinations of translational, rotational and vibrational states.

Taking logarithms of 5.3.4 gives

$$\ln q = \ln q_{\text{trans}} + \ln q_{\text{rot}} + \ln q_{\text{vib}} \qquad (5.3.5)$$

Since it is always $\ln q$ which appears in the equations for thermodynamic functions (see Sections 5.1 and 5.2) the functions U, F and S can be regarded as sums of contributions from translation rotation and vibration.

The internal energy may thus be written as

$$U = \bar{E} = NkT^2 \left(\frac{\partial \ln q}{\partial T}\right)_V$$

$$= NkT^2 \left(\frac{\partial \ln q_{\text{trans}}}{\partial T}\right)_V + \left(\frac{\partial \ln q_{\text{rot}}}{\partial T}\right)_V + \left(\frac{\partial \ln q_{\text{vib}}}{\partial T}\right)_V$$

$$= E_{\text{trans}} + E_{\text{rot}} + E_{\text{vib}} \qquad (5.3.6)$$

where

$$E_{\text{trans, rot, vib}} = NkT^2 \left(\partial \ln q_{\text{trans, rot, vib}}/\partial T\right)_V \qquad (5.3.7)$$

where the subscript "trans, rot, vib" implies either translation, rotation or vibration or any combination of them.

The free energy for an assembly of localized systems is then

$$F = -NkT \ln q$$

$$= -NkT(\ln q_{\text{trans}} + \ln q_{\text{rot}} + \ln q_{\text{vib}})$$

$$= F_{\text{trans}} + F_{\text{rot}} + F_{\text{vib}} \tag{5.3.8}$$

$$F_{\text{trans, rot, vib}} = -NkT \ln q_{\text{trans, rot, vib}} \tag{5.3.9}$$

For an assembly of non-localized systems a complication arises for we obtain

$$F = = -NkT \ln q + NkT \ln N - NkT$$

$$= -NkT \ln q_{\text{trans}} + NkT \ln N - NkT$$

$$-NkT \ln q_{\text{rot}} - NkT \ln q_{\text{vib}}$$

$$= F_{\text{trans}} + F_{\text{rot}} + F_{\text{vib}} \tag{5.3.10}$$

In identifying F_{trans}, F_{rot} and F_{vib} with terms in the second line of equation (5.3.10) the problem arises as to how the terms $NkT \ln N - NkT$ are to be assigned. Do we include them in the expression for only one of the component terms in F, or do we divide them out somehow between the three terms? We have shown in Section 5.2 that the terms $NkT \ln N - NkT$ (equivalent to $kT \ln N!$) arise entirely through the indistinguishability of the systems in an assembly of non-localized systems. The terms would therefore arise if a monatomic gas was considered which possessed only translational degrees of freedom. It is therefore necessary to include the terms in the translational contribution to the free energy. We then obtain

$$F_{\text{trans}} = -NkT \ln q_{\text{trans}} + NkT \ln N - NkT \tag{5.3.11}$$

$$F_{\text{rot, vib}} = -NkT \ln q_{\text{rot, vib}} \tag{5.3.12}$$

The appropriate equations for the components to the entropy of a gas are then

$$S_{\text{trans}} = NkT(\partial \ln q_{\text{trans}}/\partial T)_V + Nk \ln q_{\text{trans}} - Nk \ln N + Nk \tag{5.3.13}$$

$$S_{\text{rot, vib}} = NkT(\partial \ln q_{\text{rot, vib}}/\partial T)_V + Nk \ln q_{\text{rot, vib}} \tag{5.3.14}$$

The pressure of an ideal gas is given by

$$p = -(\partial F/\partial V)_T \tag{5.3.15}$$

At first sight, in view of the arguments given above, it might appear that p should have contributions from all degrees of freedom. This would be contrary to one's intuition that the pressure arises purely from the translational motion of the molecules. If we insert the appropriate expressions for F we obtain

$$p = +NkT \left(\frac{\partial}{\partial V}\right)_T \{\ln q_{\text{trans}} + \ln q_{\text{rot}} + \ln q_{\text{vib}} - \ln N + 1\}$$

$$= +NkT \left(\frac{\partial}{\partial V}\right)_T \{\ln q_{\text{trans}} + \ln q_{\text{rot}} + \ln q_{\text{vib}}\} \tag{5.3.16}$$

Equation (5.3.16) certainly suggests that there will be contributions from the three types of degrees of freedom *if* the partition functions depend upon the volume of the assembly. On examination we see that while the translational energy levels (which enter into the evaluation of q_{trans}) depend upon the volume of the container (see equations 2.1.8 and 5.2.1), the rotational and vibrational energy levels are quite independent of V. They depend only on purely molecular parameters. Thus only q_{trans} is a function of V and so the only contribution to the pressure of a gas comes from its translational motion in agreement with elementary theory.

5.4 THE STATISTICAL INTERPRETATION OF ENTROPY

We observed in Chapter 1 that the entropy of an assembly could be related to the amount of information which had to be discarded in making the transition from the detailed quantum mechanical description of an assembly to the thermodynamic description. We stated that a measure of the discarded information was the number of complexions, W, of the assembly which would be different in quantum mechanical terms but indistinguishable in thermodynamic terms, and then argued that the entropy of the assembly could be written as

$$S = k \ln W \tag{5.4.1}$$

We are now in a position to justify this relationship. In terms of the molecular partition function we have

$$S = Nk \ln q + NkT (\partial \ln q/\partial T)_V - Nk \ln N + Nk \qquad (5.4.2)$$

$$p(\text{state}, i) = \frac{\exp [-\epsilon_i/kT]}{q} \qquad (5.4.3)$$

$$q = \sum_{\text{states}} \exp [-\epsilon_1/kT] \qquad (5.4.4)$$

The terms $(-Nk \ln N + Nk)$ are included in S only for an assembly of non-localized systems. Equation (5.4.2) may be recast in terms of the probability $p(\text{state}, i)$ as follows:

$$S = Nk \left\{ \frac{\sum \exp [-\epsilon_i/kT] \times \ln q}{q} + \frac{1}{kT} \frac{\sum \epsilon_i \exp [-\epsilon_i/kT]}{q} \right\}$$

$$- Nk \ln N + Nk$$

$$= Nk \sum \{p(\text{state}, i) (\ln q + \epsilon_i/kT)\} - Nk \ln N + Nk$$

$$= -Nk \sum \{p(\text{state}, i) \ln \{p(\text{state}, i)\}\} - Nk \ln N + Nk$$
$$(5.4.5)$$

In writing the first line of equation (5.4.5) we obtain the first term by replacing $\ln q$ with its identity $\sum \exp [-\epsilon_i/kT] (1/q) \ln q$, and the second term is simply the expansion of $\partial \ln q/\partial T$. In the second and third lines we have substituted equation (5.4.3) noting that

$$\ln p(\text{state}, i) = -\epsilon_i/kT - \ln q \qquad (5.4.6)$$

For an assembly of localized systems we have, omitting the last two terms from (5.4.5),

$$S = -Nk \sum p(\text{state}, i) \ln \{p(\text{state}, i)\} \qquad (5.4.7)$$

We can now examine how equations (5.4.5) and (5.4.7) are related to (5.4.1)

Assembly of localized systems

Suppose that of the N sites N_1 are occupied by systems in quantum state 1, N_2 by systems in quantum state 2, and so on. The number of complexions, W, of an assembly with this specified distribution of quantum states amongst sites is identical to the number of ways of arranging N_1 objects of one sort, N_2 of a second sort, and so on amongst N sites, and is given by

$$W = \frac{N!}{N_1! \, N_2! \ldots} \tag{5.4.8}$$

Taking logarithms and applying Stirling's approximation gives

$$\ln W = \ln N! - \sum \ln N_i!$$
$$= N \ln N - N - \sum N_i \ln N_i + \sum N_i \tag{5.4.9}$$

Since $N = \sum N_i$, the second and last terms cancel giving

$$\ln W = N \ln N - \sum N_i \ln N_i$$
$$= -N \sum (N_i/N) \ln (N_i/N) \tag{5.4.10}$$

The ratio (N_i/N) for the most probable arrangement of quantum states amongst systems is simply the fraction of systems found in the state, i, when the assembly is at equilibrium, and is therefore identical to the probability that a system taken at random is found in state i. We therefore have

$$(N_i/N) = p(\text{state}, i) \tag{5.4.11}$$

and

$$\ln W = -N \sum p(\text{state}, i) \ln p(\text{state}, i) \tag{5.4.12}$$

Comparing (5.4.12) with (5.4.7) immediately gives equation (5.4.1)

$$S = k \ln W \tag{5.4.1}$$

The proper interpretation of W is then the number of complexions of the assembly for the most probable distribution of quantum states amongst sites.

Assembly of non-localized systems

The number of complexions for an assembly of non-localized systems has to be obtained by determining how many ways there are of placing N identical objects (the systems) into M numbered boxes (the allowable molecular quantum states). When Maxwell–Boltzmann statistics obtain, M greatly exceeds N. Since the translational quantum states are very numerous it is permissible to take them in groups, all members of any group having closely similar energies. Thus a typical group of quantum states of energy close to ϵ_i would contain M_i quantum states and N_i systems.

If the systems obeyed Bose–Einstein statistics, the number of complexions for this group can be obtained by considering a row of $M_i + N_i$ objects, M_i of one sort, which we represent by x, and N_i of another sort which we represent by o. We adopt the convention that an x possessing an o to its right represents a quantum state which is occupied by one or more systems (depending upon the number of o's). Thus the arrangement (5.4.13) implies that quantum state 1 is doubly occupied, quantum states 5 and 7 are singly occupied and so on. Quantum states 2, 3, 4, 6, 8, 9, etc. are unoccupied. Since by the convention we cannot have o's to the left of the first x, the number of

$$\begin{array}{ccccccccccccccc}
\text{x} & \text{o} & \text{o} & \text{x} & \text{x} & \text{x} & \text{x} & \text{o} & \text{x} & \text{x} & \text{o} & \text{x} & \text{x} & \text{x} & \text{x} \\
1 & & & 2 & 3 & 4 & 5 & & 6 & 7 & & 8 & 9 & 10 & 11
\end{array} \quad (5.4.13)$$

complexions is equal to the number of ways of arranging $(M_i + N_i - 1)$ objects of which $(M_i - 1)$ are of one kind and N_i of another. This is

$$W_i = \frac{(M_i + N_i - 1)!}{(M_i - 1)! \, N_i!} \quad (5.4.14)$$

When the systems obey Fermi–Dirac statistics the additional condition holds that the o's may appear only singly after an x. The number of complexions is then obtained by an alternative device. We arrange M_i x's in a row and place under them N_i o's as in 5.4.15.

$$\begin{array}{ccccccccccccc}
\text{x} & \text{x} & \text{x} & \text{x} & \text{x} & \text{x} & \text{x} & \text{x} & \text{x} & \text{x} & \text{x} & \text{x} & \text{x} \\
\text{o} & - & - & - & \text{o} & - & - & \text{o} & - & - & \text{o} & \text{o} & - \\
1 & 2 & 3 & 4 & 5 & 6 & 7 & 8 & 9 & 10 & 11 & 12 & 13
\end{array} \quad (5.4.15)$$

The number of complexions for this type of assembly is then

$$W_i = \frac{M_i!}{(M_i - N_i)! \, N_i!} \quad (5.4.16)$$

When the quantum states are sparsely occupied so that $M_i \gg N_i$, both formulae (5.4.14) and (5.4.16) can readily be shown to give the approximate result

$$W_i = (M_i)^{N_i}/N_i! \quad (5.4.17)$$

Equation (5.4.17) applies when the systems of the assembly obey

Maxwell–Boltzmann statistics. The number of complexions for the whole assembly is then:

$$W = \prod_{\text{groups}} \frac{M_i^{N_i}}{N_i!} \tag{5.4.18}$$

$$\ln W = \sum_{\text{groups}} \{N_i \ln M_i - N_i \ln N_i + N_i\} \tag{5.4.19}$$

$$= - \sum_{\text{groups}} N_i\{\ln (N_i/M_i) - 1\}$$

we now define the density of occupation of states as

$$\nu_i = N_i/M_i, \quad \nu_i \ll 1 \tag{5.4.20}$$

Thus summing over all states instead of groups of states we obtain

$$\ln W = N - \sum_{\text{states}} \nu_i \ln \nu_i \tag{5.4.21}$$

In equation (5.4.21) all reference to groups of quantum states has now disappeared, and so the arbitrary way in which the groups were defined is of no consequence.

The probability that a system taken at random is found in a state, i, for an assembly in equilibrium is given in terms of ν_i as

$$p(\text{state}, i) = \frac{\text{probable number of systems in state, } i}{\text{total number of systems in assembly}}$$

$$= \nu_i^{\text{eq}}/N \tag{5.4.22}$$

or

$$\nu_i^{\text{eq}} = Np \, (\text{state}, i) \tag{5.4.23}$$

Substituting for ν_i^{eq} into equation (5.4.21) gives finally

$$\ln W = -N \sum_{\text{states}} \{p(\text{state}, i) \ln \{p(\text{state}, i) \times N\}\} + N$$

$$= -N \sum_{\text{states}} \{p(\text{state}, i) \ln \{p(\text{state}, i)\}\} - N \ln N + N \tag{5.4.24}$$

The second line of (5.4.24) follows from the first since the sum of all $p(\text{state}, i)$ must be unity, and therefore $N \sum \{p(\text{state}, i) \ln N\} = N \ln N$. Comparing equation (5.4.24) with (5.4.5) shows that once more

$$S = k \ln W \tag{5.4.1}$$

W is to be interpreted as the number of complexions of the gaseous assembly which has a particular equilibrium distribution of occupied quantum states, this distribution being defined by a statement of the density of occupation of states for each energy.

PROBLEMS

5.1 By considering an assembly of three localized systems, each of which can exist in only two quantum states of energies ϵ_1 and ϵ_2, show that $Q = q^3$.

5.2 Devise a general proof of equation (5.1.4).

5.3 Which of the following types of systems are Bosons, and which are Fermions? (See pp. 25 and 77.)

^{3}He, ^{4}He, H, D, HD, ^{19}F, ^{13}C^{16}O, ^{12}C^{16}O, ^{12}CH$_4$, ^{12}CH$_3$D, H$_2$O, NO

5.4 Consider an assembly containing two distinguishable systems α and β which can occupy quantum state of energy ϵ_1, ϵ_2 and ϵ_3. Show that $Q = q^2$. Show that the same relationship holds however many quantum states are accessible, and particularly in the limit when α and β are most unlikely to be in the same quantum state. How does the relationship change when α and β cannot be distinguished? Which situation corresponds to the gaseous state?

5.5 Prove that equations (5.4.14) and (5.4.16) reduce to (5.4.17) when $M_i \gg N_i$.

5.6 Prove that formula (5.4.14) gives the number of ways of distributing N_i vibrational quanta amongst M_i localized one-dimensional harmonic oscillators whose energies in excess of that of the ground state are

$$\epsilon_n = nh\nu \quad \text{where } n = 0, 1, 2, 3, \text{ etc.}$$

Hence using Stirling's approximation calculate the number of complexions for the following assemblies:

Number of systems	Number of quanta
100	100
100	500
1000	1000
1000	5000

5.7 Sodium metal (atomic weight 23) has a density of about 10^3 kg m^{-3} and each sodium atom has one electron which can contribute to the electrical conductivity of the metal. To a first approximation these conductivity electrons may be regarded as behaving like molecules of a gas. Assuming that each translational state can be occupied by two electrons, and that they are filled up in order, calculate the energy of the highest occupied quantum state, and compare this with the average thermal energy of molecules of a Maxwell–Boltzmann gas, $(\frac{3}{2})kT$.

(*Note*: use the argument given at the beginning of Section 5.2 to calculate R_0, and hence the energy ϵ of the highest occupied state.)

Part II

PARTITION FUNCTIONS AND THEIR APPLICATIONS

CHAPTER 6

MOLECULAR PARTITION FUNCTIONS

The molecular partition function is defined

$$q = \sum_{\substack{\text{energy} \\ \text{levels}}} g_i \exp\left[-\epsilon_i/kT\right] \tag{6.0.1}$$

where the ϵ_i are the energies of the molecular energy levels and the g_i their degeneracies. To a first approximation the translational, rotational and vibrational motions of molecules are independent and uncoupled, and their energies may be separated. Thus by the multiplication theorem for independent modes of motion

$$q = q_{\text{trans}} q_{\text{rot}} q_{\text{vib}} \tag{6.0.2}$$

The energies of the rotational and vibrational states are not, however, generally independent of the electronic states of molecules and we cannot include the electronic partition function q_{el} in this product except in special cases. Instead we must write generally when electronic quantum states are important

$$q = q_{\text{trans}} \sum \left\{ g_{j(\text{el})} \exp\left[-\frac{\epsilon_{j(\text{el})}}{kT}\right] q_{j(\text{rot})} q_{j(\text{vib})} \right\} \tag{6.0.3}$$

where $q_{j(\text{rot})}$ and $q_{j(\text{vib})}$ are the rotational and vibrational partition functions for the jth electronic energy level whose degeneracy is $g_{j(\text{el})}$.

For the majority of molecules only the lowest electronic energy level is important because of the large differences in the energies of these levels, and negligible error is made if equation (6.0.2) is used for the molecular partition function measured from the lowest electronic energy state of the molecule.

6.1 THE TRANSLATIONAL PARTITION FUNCTION

The energy levels for a particle of mass m in a field free rectangular box of sides a, b and c are

$$\epsilon_{\text{trans}} = \frac{h^2}{8m} \left\{ \frac{j^2}{a^2} + \frac{l^2}{b^2} + \frac{n^2}{c^2} \right\} \tag{6.1.1}$$

95

and are non-degenerate. The partition function is then given by

$$q_{\text{trans}} = \sum_{j,\,l,\,n} \exp\left[\frac{-h^2}{8mkT}\left(\frac{j^2}{a^2} + \frac{l^2}{b^2} + \frac{n^2}{c^2}\right)\right] \tag{6.1.2}$$

where the triple summation is to be taken over all integral values of j, l and n from one to infinity. Since the motions in the three directions are independent the triple summation can be written as a product of three summations. This arises by the converse of the multiplication theorem. Thus

$$q_{\text{trans}} = \sum_{j} \exp\left[-\frac{h^2 j^2}{8ma^2kT}\right] \times \sum_{l} \exp\left[-\frac{h^2 l^2}{8mb^2kT}\right]$$

$$\times \sum_{n} \exp\left[-\frac{h^2 n^2}{8mc^2kT}\right] \tag{6.1.3}$$

As was shown in Section 5.2 the energy levels for translation are very closely spaced compared to kT. The summations can then be replaced by integrations in the same way as the summation for $\ln N!$. To a high degree of accuracy

$$q_{\text{trans}} = \int_0^\infty \exp\left[-\frac{h^2 j^2}{8ma^2kT}\right] dj \times \int_0^\infty \exp\left[-\frac{h^2 l^2}{8mb^2kT}\right] dl$$

$$\times \int_0^\infty \exp\left[-\frac{h^2 n^2}{8mc^2kT}\right] dn \tag{6.1.4}$$

The three integrals appearing in (6.1.4) are identical, apart from the constants a, b and c. They are of the form $\int_0^\infty \exp\left[-x^2/2\sigma^2\right] dx$, the integral of the Gaussian error curve. The value of this integral which is evaluated below is given by (6.1.5)

$$I = \int_{-\infty}^\infty \exp\left[-\frac{x^2}{2\sigma^2}\right] dx = 2\int_0^\infty \exp\left[-\frac{x^2}{2\sigma^2}\right] dx = (2\pi\sigma^2)^{1/2} \tag{6.1.5}$$

For the first integral in (6.1.4), $\sigma^2 = 4ma^2kT/h^2$ and the integral is $(2\pi mkT)^{1/2}(a/h)$. The translational partition function for motion in a rectangular box of sides a, b and c is thus

$$q_{\text{trans}} = \left\{\frac{2\pi mkT}{h^2}\right\}^{3/2} \times (abc) \tag{6.1.6}$$

The factor (abc) is the volume of the box, V, giving the result which may be generalized to any shape of box

$$q_{\text{trans}} = \left(\frac{2\pi mkT}{h^2}\right)^{3/2} \times V = q^{\circ}_{\text{trans}} \times V \qquad (6.1.7)$$

where q°_{trans} is the translational partition function *per unit volume*.

The integral I of (6.1.5) may be obtained by a mathematical trick. We start with the square of the integral which may be written

$$I^2 = \int_{-\infty}^{\infty} \exp\left[-\frac{x^2}{2\sigma^2}\right] dx \int_{-\infty}^{\infty} \exp\left[-\frac{y^2}{2\sigma^2}\right] dy$$

$$= \iint_{-\infty}^{\infty} \exp\left[-\frac{x^2 + y^2}{2\sigma^2}\right] dx \, dy \qquad (6.1.8)$$

Since the equation $x^2 + y^2 = r^2$ describes a circle round the origin, the integrand has the same value for all points on the circumference of any such circle. As shown in Figure 6.1, the integration may thus be

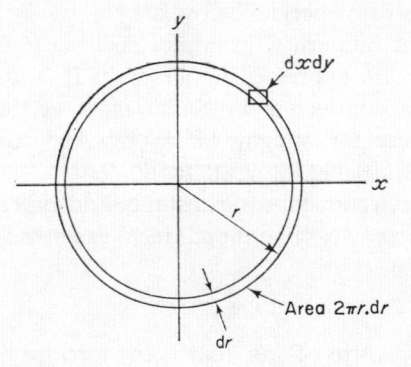

Figure 6.1 Replacement of an area element $dx \, dy$ by a ring element $2r \, dr$.

performed in a series of circular rings by replacing the rectangular area element $dx \, dy$ by a circular ring element of area $2\pi r \, dr$. The integral of (6.1.8) thus becomes

$$I^2 = \int_0^{\infty} \exp\left[-\frac{r^2}{2\sigma^2}\right] 2\pi r \, dr = \pi \int_0^{\infty} \exp\left[-\frac{r^2}{2\sigma^2}\right] d(r^2) = 2\pi\sigma^2$$

$$(6.1.9)$$

In (6.1.9) the limits of integration must now be $0 < r < \infty$ to cover the complete area of the plane of integration. The result of (6.1.9) is the same as given in (6.1.5).

The translational partition function given by equation (6.1.7) is, for all practical purposes, an extremely large number. For example for benzene (molecular weight, $MW = 78 \cdot 0$) in a volume of 1 m^3 at 25°C (298 K) we obtain

$$q_{\text{trans}} = \left\{\frac{2\pi MWkT}{h^2 L_A}\right\}^{3/2} V$$

$$= \left\{\frac{2 \times 3\cdot142 \times 78\cdot0 \times 1\cdot380 \times 10^{-23} \times 298}{6\cdot626^2 \times 10^{-68} \times 6\cdot022 \times 10^{26}}\right\}^{3/2} \times 1\cdot00$$

$$= 6\cdot66 \times 10^{32} \qquad (6.1.10)$$

The partition function is always a pure number and is generally of the same order as the number of quantum states with energies below the typical "thermal energy" kT. Reference to equation (5.2.4) shows that the translational partition function is $3\sqrt{\pi}/4 = 1\cdot33$ times the number of quantum states with energies below kT.

In working out numerical examples such as (6.1.10) when fundamental constants are expressed in S.I. units it is important to notice that if a formula involves a molecular mass, m, this is obtained by dividing the molecular weight, MW, by Avogadro's number, L_A, expressed in molecules per kilogram mole.

Under certain circumstances substances adsorbed on the surfaces of liquids behave as two-dimensional ideal gases and obey an equation of state of the form

$$pA = NkT \qquad (6.1.11)$$

where p = line pressure of gas, that is the force per unit length of the film edge, and A = film area. For this two-dimensional state of matter the appropriate partition function is

$$q_{\text{trans(2-D)}} = (2\pi mkT/h^2) \times A \qquad (6.1.12)$$

In a similar way some species adsorbed by molecular sieves (synthetic Zeolites) can be regarded to a first approximation as one-dimensional gases. The molecules of such a "gas" are arranged in rows in the long tubular cavities within the crystal lattice. The one-dimensional equation of state is then

$$pL = NkT \qquad (6.1.13)$$

where p is the force exerted by one such row of molecules, and L is the length of the row containing N molecules. The appropriate partition function is then

$$q_{\text{trans(1-D)}} = (2\pi mkT/h^2)^{1/2} \times L \qquad (6.1.14)$$

6.2   THE ROTATIONAL PARTITION FUNCTIONS

The rotational motion of molecules may formally be regarded as one-, two- or three-dimensional.

The free internal rotation of one part of a molecule against another part is an example of one-dimensional rotational motion. It occurs in molecules such as dimethyl cadmium, and dimethyl acetylene where the two methyl groups are so far apart that there is virtually no inter-action between them and no hindrance to rotation of one group relative to the other around the molecular axis. In molecules like ethane there is strong non-bonding interaction between the two ends of the molecule and rotation is severely hindered. Indeed around room temperature the majority of ethane molecules undergo torsional oscil-lation about the staggered position. Only those few molecules which contain quite large amounts of energy in the torsional mode can under-go rotation of one methyl group relative to the other. The angular potential function for restricted internal rotation is often assumed to be of the cosine form given in equation (6.2.1) and shown in Figure 6.2

$$V(\phi) = (V_0/2)(1 - \cos n\phi) = V_0 \sin^2(n\phi/2) \qquad (6.2.1)$$

where ϕ is the angular coordinate of one of the methyl groups relative to the other, V_0 is the "height" of the potential barrier, and n the num-ber of minima in the potential function for one complete revolution; $n = 3$ for two methyl groups. For ethane the height of the barrier to internal rotation is about 13 kJ mol^{-1}, while in dimethyl acetylene it is probably below 1 kJ mol^{-1}.

Two-dimensional rotators include all diatomic molecules, and all linear polyatomic molecules, for example, O_2, HCl, CO_2, C_2H_2, but not bent polyatomic molecules such as H_2O, H_2O_2, O_3, etc. even if they are planar. According to classical mechanics the rotational motion of any linear body can be resolved into component rotations about two mutually perpendicular axes (see Section 7.2) both of which are per-pendicular to the molecular axis and pass through its centre of mass.

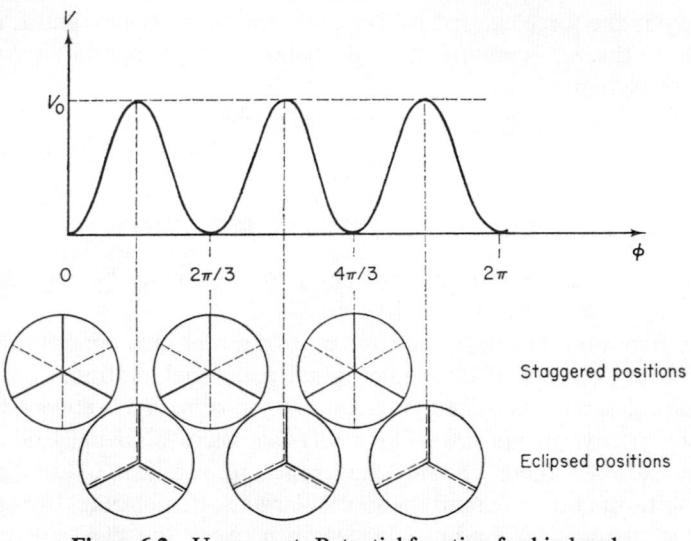

Figure 6.2 Upper part: Potential function for hindered internal rotation, equation (6.2.1) lower part: configurations of ethane for various ϕ as seen end on.

When applying this to linear molecules it is sometimes suggested that there ought to be a third rotational degree of freedom representing rotation around the molecular axis, for the atoms of a molecule certainly cannot be truly regarded as point masses, and one might expect that linear molecules would possess a moment of inertia about such an axis. Although such an argument would be valid in classical terms, it has no meaning in quantum mechanical terms.

All non-linear polyatomic molecules behave as three-dimensional rotators. They possess three moments of inertia for rotation about the three so-called principal axes of rotation. These three axes are mutually perpendicular and pass through the centre of mass of the molecule. They can often be identified by inspection if the molecules have a high degree of symmetry. In benzene, for example, one axis must pass through the centre of the ring and be perpendicular to it. The other two axes must be in the plane of the ring, one through two opposite carbon atoms, and the other through the centres of two sides. Examples showing the principal axes for some symmetrical molecules are given in Figure 6.3. The principal axes are unique for most molecules, and are those axes about which it would spin without exerting any twisting force

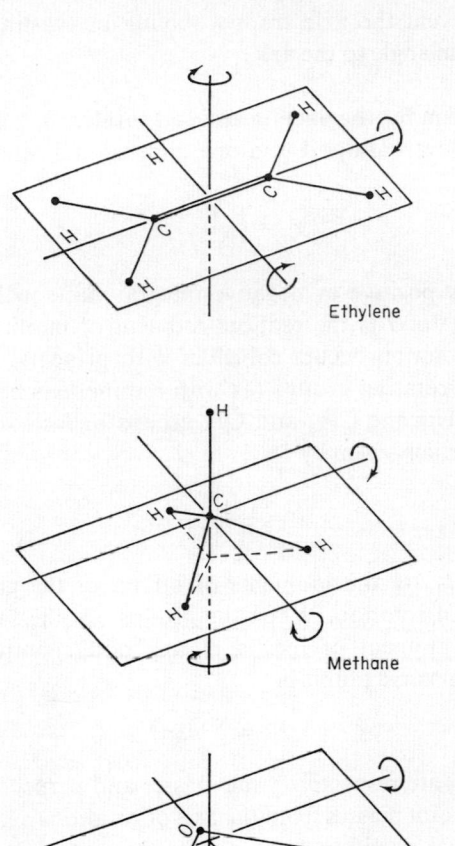

Figure 6.3 Principal axes of rotation of ethylene, methane and water molecules.

on the axis. An example of an axis which is *not* a principal axis is the line of the axle of a buckled wheel. When such a wheel rotates it is difficult to prevent the axle from wobbling or twisting. The principal axis is then at an angle to the axle.

Partition function for the one-dimensional rotator

The energy levels allowed to a one-dimensional rotator (Section 2.1) are

$$\epsilon_j = \left\{\frac{h^2}{8\pi^2 I}\right\} j^2 \tag{6.2.2}$$

j may take any positive or negative integral value including zero. For an internal rotator I is the reduced moment of inertia. When the axis about which rotation occurs coincides with principal axes of rotation of the counter-rotating groups (as with the methyls in dimethyl acetylene, but not with the CH_3 and OH groups in methanol) the reduced moment of inertia is given by

$$I = \frac{I_1 I_2}{I_1 + I_2} \tag{6.2.3}$$

where I_1 and I_2 are the moments of inertia of the groups about the common axis of rotation. When the groups are the same $I_1 = I_2$ and $I = I_1/2$. The moment of inertia of any body composed of a rigid framework of massive points is

$$I_1 = \sum m_i r_i^2 \tag{6.2.4}$$

where m_i and r_i are respectively the masses and perpendicular distances of each of the point masses from the axis of rotation.

The rotational partition function for free internal rotation is thus given by

$$q_{\text{rot(1-D)}} = \sum_{j=0,\,\pm 1,\,\pm 2,\ldots}^{\pm\infty} \exp\left[-\frac{h^2 j^2}{8\pi^2 IkT}\right] \tag{6.2.5}$$

If the energy levels are sufficiently closely spaced compared to kT we may replace the summation by an integration as was done in evaluating q_{trans}. For the moment we assume this to be allowable and obtain

$$q_{\text{rot(1-D)}} = \int_{-\infty}^{\infty} \exp\left[-j^2 \frac{\theta_r}{T}\right] dj = \left(\frac{\pi T}{\theta_r}\right)^{1/2} = \sqrt{\pi}\left(\frac{8\pi^2 IkT}{h^2}\right)^{1/2} \tag{6.2.6}$$

In evaluating the integral it is convenient to replace the constant $h^2/8\pi^2 Ik$ by θ_r. θ_r has the dimensions of temperature and is called the "characteristic rotational temperature" for the internal rotation. The value of θ_r/T is seen from equation (6.2.2) to give an indication of whether or not the rotational energy levels are close with respect to kT. We may therefore expect the approximation (6.2.6) to hold well if θ_r is much less than T, but for the accuracy of (6.2.6) to weaken as T approaches θ_r.

Example 6.1 The partition function for the internal rotation in dimethyl acetylene.

The structure of dimethyl acetylene is shown in Figure 6.4. The C—H

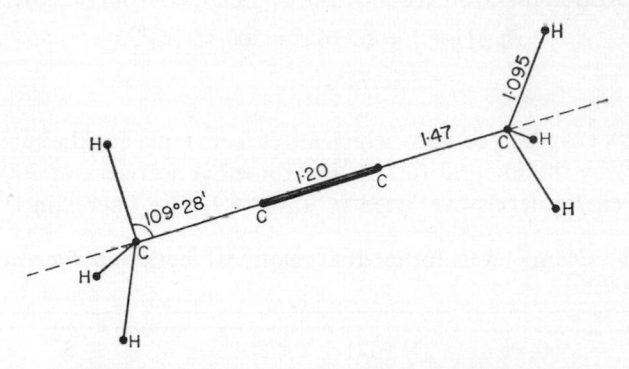

Figure 6.4 Dimensions of dimethyl acetylene molecule: bond lengths in Angstrom units. 1 Å = 10^{-10} m.

bond length is $1{\cdot}095 \times 10^{-10}$ m, that is $1{\cdot}095$ Å, and the angle between the C—H bonds and the axis of rotation (the line joining the four C atoms) is $70° \, 32'$ (the supplement of $109° \, 28'$). The perpendicular distance of each H atom from the axis of rotation is then

$$r_i = 1{\cdot}095 \sin 70° \, 32' \text{ Å} = 1{\cdot}032 \text{ Å}$$

and moment of inertia for the CH_3 group is

$$I_{Me} = (12{\cdot}01 \times 0^2 + 3 \times 1{\cdot}008 \times 1{\cdot}032^2) \text{ amu Å}^2$$
$$= 3{\cdot}210 \text{ amu Å}^2$$
$$= 3{\cdot}210 \times 10^{-20}/(6{\cdot}023 \times 10^{26}) \text{ kg m}^2$$
$$= 5{\cdot}32 \times 10^{-47} \text{ kg m}^2$$

In practice it is usually simplest to evaluate moments of inertia initially in amu Å², and convert later to kg m² if necessary.

The reduced moment of inertia is then

$$I = I_{Me}/2 = 1{\cdot}605 \text{ amu Å}^2 = 2{\cdot}66 \times 10^{-47} \text{ kg m}^2$$

The characteristic rotational temperature, θ_r is then

$$\theta_r = h^2/8\pi^2 Ik$$

$$= \frac{(6{\cdot}626 \times 10^{-34})^2}{8 \times 3{\cdot}142^2 \times 2{\cdot}66 \times 10^{-47} \times 1{\cdot}381 \times 10^{-23}}$$

$$= 15{\cdot}26 \text{ K}$$

The partition function at 300 K is then obtained from (6.2.6) as

$$q_{rot}(1\text{-}D) = (3{\cdot}142 \times 300/15{\cdot}16)^{1/2}$$

$$= 7{\cdot}90$$

We now examine the error which arises from replacing the summation of (6.2.5) by the integral (6.2.6). For dimethyl acetylene at 300 K, the allowed energy levels are listed in Table 6.1. The figures in the third

Table 6.1   Energy levels for internal rotational motion in dimethyl acetylene at 300 K

j	0	1	2	3	4	5	6
$10^{22}\epsilon_j/J$	0	2·09	8·4	18·8	33·5	52·3	75·2
ϵ_j/kT	0	0·051	0·202	0·455	0·81	1·26	1·82

row show that the internal rotational energy levels are *not* in fact close together compared to kT, and that the assumption that they are, might lead to considerable error in q. Figure 6.5 shows quantitatively how the sum and integral differ. The area under the stepped curve is the value of the summation and that under the smooth curve the value of the integral. Although the shaded areas largely cancel, numerical evaluation of the sum and integral for various values of θ_r/T shows that the sum always exceeds the integral by a small amount. However, the error of the integral is only about 0·1% when $\theta_r/T = 1{\cdot}2$, and is negligible when θ_r/T is less than unity. Thus with dimethyl acetylene, which has the lowest reduced moment of inertia of any internal rotator, and for which $\theta_r/T = 0{\cdot}051$ at 300 K, equation (6.2.6) is highly accurate.

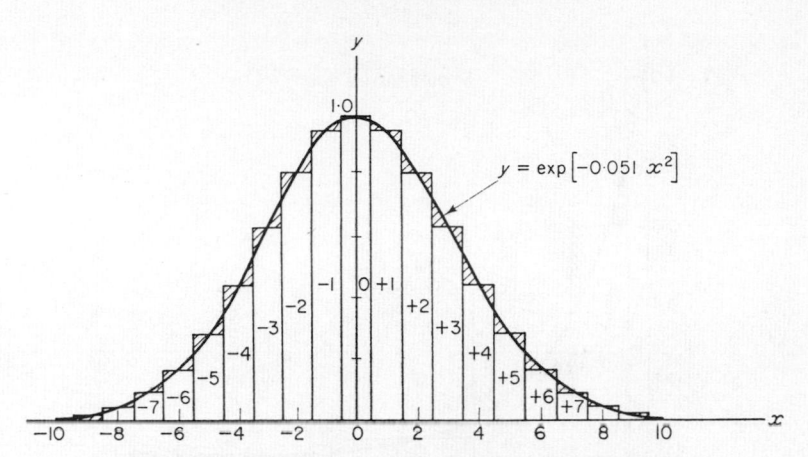

Figure 6.5 Comparison of integral and summation forms of the partition function for one-dimensional rotation. Integral form, smooth curve; summation form, stepped curve. Substance, dimethylacetylene at 300 K.

Partition function for the two-dimensional rotator

The energy levels for a two-dimensional rotator are

$$\epsilon_j = (h^2/8\pi^2 I)j(j+1)$$
$$= j(j+1)k\theta_r \qquad (6.2.7)$$

where $\theta_r = h^2/8\pi^2 Ik$ is the "characteristic rotational temperature". The degeneracy of each level is $(2j+1)$. The partition function is

$$q_{\text{rot(2-D)}} = \sum_{j=0}^{\infty} (2j+1)\exp\left[-j(j+1)\frac{\theta_r}{T}\right] \qquad (6.2.8)$$

Assuming that the exact summation may be replaced by an integration we obtain

$$q_{\text{rot(2-D)}} = \int_0^\infty (2j+1)\exp\left[-j(j+1)\frac{\theta_r}{T}\right]dj$$
$$= \int_0^\infty \exp\left[-j(j+1)\frac{\theta_r}{T}\right]d\{j(j+1)\}$$
$$= \int_0^\infty \exp\left[-\left(\frac{\theta_r}{T}\right)x\right]dx$$
$$= \frac{T}{\theta_r} = \frac{8\pi^2 IkT}{h^2} \qquad (6.2.9)$$

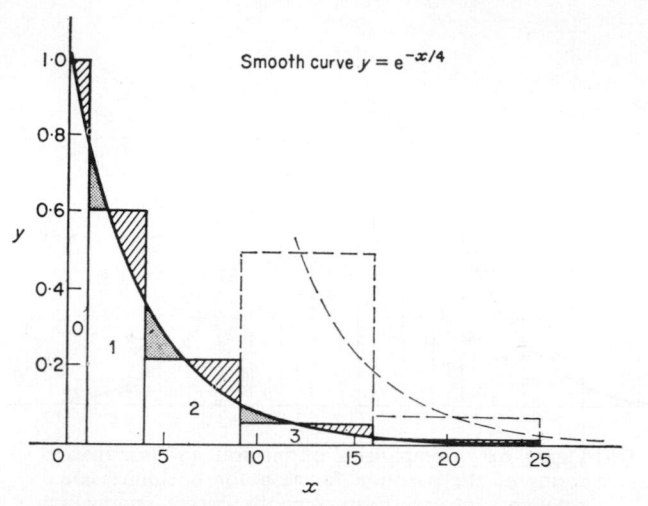

Figure 6.6 Comparison of integral and summation forms of partition function for two-dimensional rotation. Integral form, smooth curve; summation form, stepped curve. Substance, HD at 263 K.

The error which arises in replacing the exact summation by the integration is illustrated in Figure 6.6. The percentage error as a function of θ_r/T being given in Figure 6.7. In Figure 6.6 the smooth curve is the exponential decay curve

$$y = \exp\left[-(\theta_r/T)x\right], \quad \theta_r/T = 0.25 \tag{6.2.10}$$

The area under this curve, as noted from the third line of equation (6.2.9), is the value of the integral. The area under the stepped curve gives the true value of q_{rot}. Each rectangular step corresponds to one term in the summation (6.2.8). The height of each step is $\exp\left[-j(j+1)(\theta_r/T)\right]$, and the width is $(2j+1)$. The smooth curve intersects the stepped curve at $x = j(j+1)$, which is j units from the start and $(j+1)$ units from the end of the step. It is easy to see that the area under the stepped curve is greater than that under the smooth curve in spite of a partial cancelling out of the dotted and shaded areas. Again the integrated form underestimates q_{rot}. A good approximation to the true partition function is given by Mulholland's equation

$$q_{\text{rot(2-D)}} = \frac{T}{\theta_r}\left\{1 + \frac{1}{3}\frac{\theta_r}{T} + \frac{1}{15}\left(\frac{\theta_r}{T}\right)^2\right\} \tag{6.2.11}$$

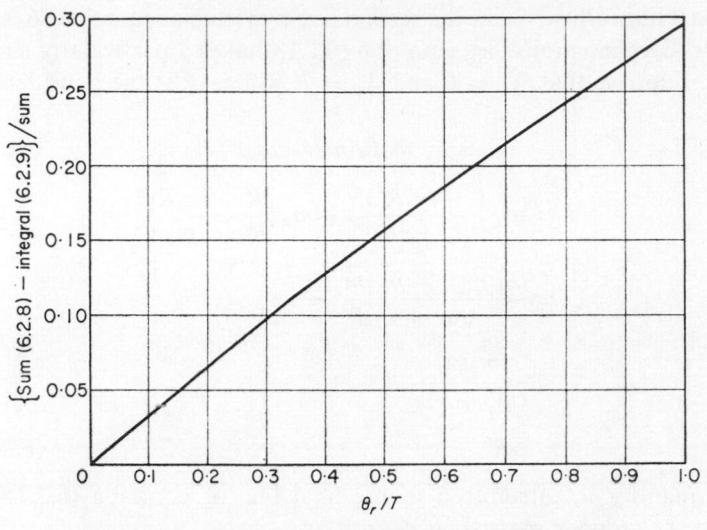

Figure 6.7 Fractional error for approximation to rotational partition function for two-dimensional rotation.

Formula (6.2.11) is 1% low when $\theta_r/T = 0.9$, while formula 6.2.9 is 1% low when $\theta_r/T = 0.03$. Generally the latter condition holds and equation (6.2.10) is almost universally used, but it is important to examine the conditions when θ_r/T is likely to exceed 0.03. First we digress briefly on the evaluation of the moments of inertia of linear molecules.

The moment of inertia of any linear molecule is given by equation (6.2.4) if the r_i are the distances of the nuclei from the centre of mass of the molecule. The position of the centre of mass is found by the principle of moments. If $X_1, X_2, \ldots$ are the coordinates of the masses $m_1, m_2, \ldots$ measured from some arbitrary origin on the line of the nuclei, the coordinate of the centre of mass, $\bar{X}$, is obtained from the equation

$$0 = \sum m_i(X_i - \bar{X}) \quad \text{or} \quad \bar{X} = \frac{\sum m_i X_i}{\sum m_i} = \frac{\sum m_i X_i}{M} \quad (6.2.12)$$

where M is the total mass of the molecule. The general formula for the moment of inertia is then

$$\begin{aligned}
I &= \sum m_i(X_i - \bar{X})^2 \\
&= \sum m_i X_i^2 - 2\bar{X}\sum m_i X_i + \bar{X}^2 \sum m_i \\
&= \sum m_i X_i^2 - M\bar{X}^2 \quad (6.2.13)
\end{aligned}$$

The last line follows from the second line by virtue of equation (6.2.12).

For diatomic molecules equation (6.2.13) takes a particularly simple form. Suppose that $X_1 = 0$ and $X_2 = R$, where R is the bond length. Then

$$\bar{X} = m_2 R/(m_1 + m_2) \tag{6.2.14}$$

$$I = m_1 \left\{ \frac{-m_2 R}{m_1 + m_2} \right\}^2 + m_2 \left\{ \frac{R - m_2 R}{m_1 + m_2} \right\}^2$$

$$= \frac{(m_1 m_2{}^2 + m_1{}^2 m_2) R^2}{(m_1 + m_2)^2}$$

$$= \frac{m_1 m_2}{(m_1 + m_2)} R^2$$

$$= \mu R^2 \tag{6.2.15}$$

The quantity μ, introduced in the final line of equation (6.2.15), is called the reduced mass. It is defined as

$$\text{Reduced mass, } \mu = m_1 m_2/(m_1 + m_2) \tag{6.2.16}$$

The diatomic rotator can thus be treated as a single point mass, μ, rotating about a centre a distance R away.

Table 6.2 lists the rotational and vibrational properties of a number of diatomic and linear polyatomic molecules. For all molecules listed except hydrogen, θ_r/T at the boiling point is low, but for diatomic molecules which contain H atoms and so by formula (6.2.15) have low moments of inertia θ_r/T may be above 0·03 at the boiling point. Nevertheless even for these molecules expression (6.2.9) introduces little error at higher temperatures. Hydrogen, however, presents a special case, not only because of its low moment of inertia but because of its extremely low boiling point. The rotational properties of hydrogen are considered later in Chapter 9.

Partition function for the three-dimensional rotator

The evaluation of the partition function for the three-dimensional rotator is beyond the scope of this book and we simply quote the result:

$$q_{\text{rot(3-D)}} = \sqrt{\pi}(8\pi^2 kT/h^2)^{3/2}(ABC)^{1/2} \tag{6.2.17}$$

A, B and C are the moments of inertia for rotation about the three principal axes of rotation. The result is obtained formally by multi-

Table 6.2 Rotational properties of linear molecules

	T_{bp}/K	$10^{-10}R/m$	$I/amu\ Å^2$	θ_r/K	ω/cm^{-1}	θ_v/K^a
H_2	14·02	0·7415	0·2771	87·53	4405	6338
N_2	77·4	1·0976	8·437	2·874	2360	3395
O_2	89·8	1·2074	11·663	2·080	1580	2274
Cl_2	239·1	1·988	70·06	0·346	565	813
Br_2	332	2·283	208·2	0·1165	323	465
I_2	458	2·666	451·0	0·0538	215	308
HCl	188·1	1·2744	1·593	15·24	2990	4302
HBr	206	1·408	1·973	12·29	2650	3812
HI	238	1·608	2·586	9·38	2310	3323
$BrCl$	278	2·138	112·3	0·216	519	747
ICl	371	2·307	147·5	0·165	384	553
CO	81·7	1·1282	8·733	2·777	2170	3122
NO	121·4	1·1506	9·888	2·453	1904	2739
CO_2	194	1·1600	43·06	0·5644	b	b
C_2H_2	170	1·061 CH 1·202 CC	15·22	1·593	c	c

^{a} For definition of θ_v see p. 125 equation (6.3.15).
^{b} For CO_2 the fundamental frequencies and θ_v/K values are: 1351, 1944; 2396, 3447; 672 (degeneracy 2), 967 (2).
^{c} For C_2H_2 the fundamental frequencies and θ_v/K values are: 3374, 4855; 1974, 2480; 3287, 4729; 612(2), 881(2); 729(2), 1049(2).

plying the result for the one-dimensional rotator by that for the two-dimensional rotator having replaced I in the latter equation by $(AB)^{1/2}$.

Formula (6.2.17) applies to all non-linear polyatomic molecules including such planar molecules as H_2O, C_2H_4, etc. It is a semi-classical partition function and embodies the same sort of approximation as do equations (6.2.6) and (6.2.10). However, the moments of inertia of poly-atomic molecules are usually sufficiently high that any errors involved are insignificant. The only exception to this is methane whose boiling point is low, and whose moments of inertia (see below) are also low.

Molecules for which A, B and C are all different are called asymmetric tops, those with $A = B$ (for example NH_3) are called symmetric tops, and those with $A = B = C$ (for example methane) are called spherical tops.

The three moments of inertia about the principal axes taken as the x, y and z cartesian axes are given by

$$A = I_{xx} = \sum m_i(y_i^2 + z_i^2)$$
$$B = I_{yy} = \sum m_i (z_i^2 + x_i^2)$$
$$C = I_{zz} = \sum m_i(x_i^2 + y_i^2) \qquad (6.2.18)$$

The factors in brackets, such as $(y_i^2 + z_i^2)$ are, of course, the squares of the perpendicular distances of the nuclei from the axes.

When the molecule is planar, all the z_i are zero and $I_{zz} = I_{xx} + I_{yy}$. Thus $ABC = I_{xx}I_{yy}(I_{xx} + I_{yy})$.

Example 6.2 ABC and the rotational partition function for ethylene.

Ethylene is planar. Its dimensions are shown in Figure 6.8. The principal axes may be determined by inspection. The x-axis is taken

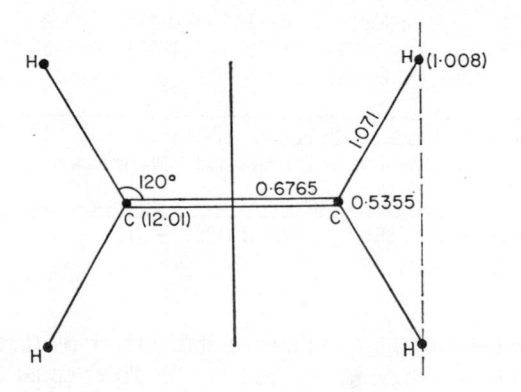

Figure 6.8 Dimensions of ethylene molecule. Bond lengths in Å.

along the C—C bond, and the z-axis perpendicular to the plane of the molecule. The coordinates and masses of the atoms of the molecules and the products required for evaluation of the moments of inertia are listed in Table 6.3, from which we obtain

$$ABC = 1200 \text{ amu}^3 \text{ Å}^6$$
$$= 1200 \times 10^{-60} (6 \cdot 0225 \times 10^{26})^{-3} \text{ kg}^3 \text{ m}^6$$
$$= 5 \cdot 49 \times 10^{-138} \text{ kg}^3 \text{ m}^6$$

Table 6.3 Coordinates of atoms in ethylene

Atom	m_i/amu	x_i/Å	y_i/Å	z_i/Å	$m_i x_i^2$	$m_i y_i^2$	$m_i z_i^2$
H_a	1·008	−1·212	−0·9275	0	1·484	0·867	0
H_b	1·008	−1·212	0·9275	0	1·484	0·867	0
H_c	1·008	1·212	0·9275	0	1·484	0·867	0
H_d	1·008	1·212	−0·9275	0	1·484	0·867	0
C_a	12·01	−0·6765	0	0	5·50	0	0
C_b	12·01	0·6765	0	0	5·50	0	0
				Sums	16·94	3·47	0

Moments of inertia
$$A = I_{xx} = 3·47 + 0 = 3·47 \text{ amu Å}^2$$
$$B = I_{yy} = 16·94 + 0 = 16·94 \text{ amu Å}^2$$
$$C = I_{zz} = 16·94 + 3·47 = 20·41 \text{ amu Å}^6$$

Moment of inertia product
$$ABC = 3·47 \times 16·94 \times 20·41 = 1199 \text{ amu}^3 \text{ Å}^6$$

The rotational partition function for ethylene ignoring nuclear symmetry considerations, which are discussed below, is then

$$q_{\text{rot}(C_2H_4)} = \sqrt{\pi}(8\pi^2 k/h^2)^{3/2} \times T^{3/2} \times (ABC)^{1/2}$$
$$= 21·931 \times 10^{67} \times (300)^{3/2} \times (5·49 \times 10^{-138})^{1/2}$$
$$= 2·67 \times 10^3$$

Example 6.3 The moments of inertia and rotational partition function for methane.

Methane is a highly symmetrical molecule, and, as shown here, is

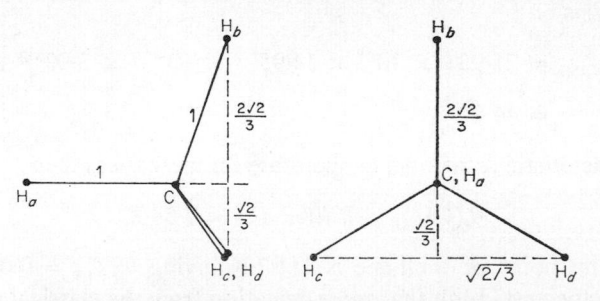

Figure 6.9 Dimensions of methane molecule with C—H bond lengths given as one arbitrary unit. Actual C—H bond length is 1·091 Å.

spherical top. The C atom is at the centre of mass and so contributes nothing to the moment of inertia. Figure 6.9 gives the geometry of the molecule. For simplicity the C—H bond length has been given as one arbitrary unit. Its actual length is 1.091 Å.

Taking one C—H bond as the x-axis, and another in the x, y-plane, the coordinates of the four atoms and the associated products in arbitrary units are

Atom	x_i	y_i	z_i	x_i^2	y_i^2	z_i^2
H_a	-1	0	0	1	0	0
H_b	$1/3$	$2\sqrt{2}/3$	0	$1/9$	$8/9$	0
H_c	$1/3$	$-\sqrt{2}/3$	$\sqrt{(2/3)}$	$1/9$	$2/9$	$2/3$
H_d	$1/3$	$-\sqrt{2}/3$	$-\sqrt{(2/3)}$	$1/9$	$2/9$	$2/3$
			Sums	$4/3$	$4/3$	$4/3$

The three moments of inertia are evidently identical. Introducing the bond length in Å and the mass of the H atom in amu gives

$$A = B = C = (4/3 + 4/3) \times 1.091^2 \times 1.008 \text{ amu Å}^2$$

$$= 3.197 \text{ amu Å}$$

$$ABC = 32.7 \text{ amu}^3 \text{ Å}^6$$

$$= 14.95 \times 10^{-140} \text{ kg}^3 \text{ m}^6$$

The partition function for methane at 300 K, again ignoring nuclear symmetry, is

$$q_{rot(CH_4)} = 21.931 \times 10^{67} \times 14.95^{1/2} \times 10^{-70} \times 3.00^{3/2} \times 10^3$$

$$= 44.2$$

The characteristic rotational temperature for methane is then

$$\theta_r(CH_4) = h^2/(8\pi^2 Ak) = 7.58 \text{ K}$$

The boiling point of methane is 111.7 K giving $\theta_r/T_{bp} = 0.067$. This value is sufficiently high that some deviation from the classical partition function may be expected at around this temperature.

For many molecules the principal axes of rotation cannot be found by inspection. A few simple examples are HDO, CH_3OH, CH_2FCl, NH_2F. It is then necessary to use a more complicated formula to obtain ABC:

$$ABC = \begin{vmatrix} I_{xx} & -I_{xy} & -I_{xz} \\ -I_{xy} & I_{yy} & -I_{yz} \\ -I_{xz} & -I_{yz} & I_{zz} \end{vmatrix} \qquad (6.2.19)$$

I_{xx}, I_{yy} and I_{zz} are given by formula (6.2.18) where the coordinates are taken relative to any set of mutually perpendicular axes passing through the centre of mass. The other quantities in the determinant, called the products of inertia, are evaluated by equation (6.2.20)

$$I_{xy} = \sum m_i x_i y_i = I_{yx}$$

$$I_{yz} = \sum m_i y_i z_i = I_{zy} \qquad (6.2.20)$$

$$I_{zx} = \sum m_i x_i z_i = I_{xz}$$

If the axes are chosen to be the principal axes the products of inertia vanish. For all molecules with a mirror plane one of the principal axes must be the line through the centre of mass perpendicular to the mirror plane. If this axis is the z-axis then the products of inertia I_{yz} and I_{zx} are zero and the determinant simplifies to

$$ABC = I_{zz}(I_{xx}I_{yy} - I_{xy}^2) \qquad (6.2.21)$$

In unsymmetrical molecules neither the centre of mass nor the principal axes can be found by inspection, and a calculation procedure is desirable which enables any arbitrary set of cartesian axes to be used. Suppose that x, y and z are coordinates relative to a particular set of Cartesian axes through the centre of mass, and X, Y, Z are coordinates relative to a set of parallel axes through some arbitrary origin. Then as shown in equation (6.2.13).

$$\sum m_i x_i^2 = \sum m_i X_i^2 - (\sum m_i X_i)^2 / M \qquad (6.2.22)$$

with similar equations involving y and z. Similar equations apply to the products of inertia:

$$\sum m_i x_i y_i = \sum m_i X_i Y_i - (\sum m_i X_i)(\sum m_i Y_i)/M \qquad (6.2.23)$$

To evaluate ABC systematically by equation (6.2.19) we therefore tabulate the following groups of quantities:

(a) Atom, m_i;   (b) X_i, Y_i, Z_i;   (c) m_iX_i, m_iY_i, m_iZ_i;

(d) $m_iX_i{}^2$, $m_iY_i{}^2$, $m_iZ_i{}^2$;   (e) $m_iX_iY_i$, $m_iY_iZ_i$, $m_iZ_iX_i$

The m_i and the quantities in groups (c), (d) and (e) are then summed to give the moments and products of inertia about the centre of mass. The procedure is readily programed on a computer or calculator.

Example 6.4   The moment of inertia product and partition function for methanol.

The structure and dimensions of the methanol molecule are shown in Figure 6.10. It is assumed that the stable configuration of the mole-

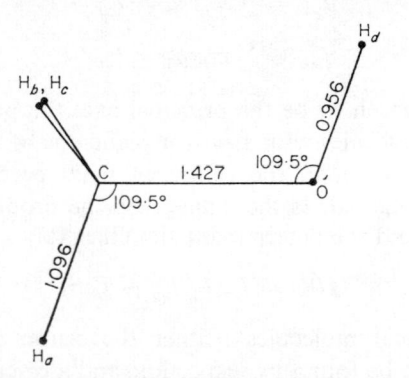

Figure 6.10   Dimensions of methanol molecule. Bond lengths in Å.

cule has the atoms H—C—O—H in a plane with the H atoms trans to the C—O bond. The other two H atoms are then above and below the plane. The plane is a mirror plane, and we take the Z-axis perpendicular to it and through the C atom which is taken as the arbitrary origin in the X, Y, Z frame of reference. The X-axis is taken as the line along the C—O bond.

The geometry of the tetrahedral grouping shown for methane in Figure 6.9 enables the coordinates of the H atoms to be readily obtained. The fourteen quantities required for calculation of ABC are

Table 6.4 Parameters for evaluation of moment of inertia product of methanol

Atom	m_i	X_i	Y_i	Z_i	$m_i X_i$	$m_i Y_i$	$m_i Z_i$
H_a	1·008	−0·366	−1·032	0	−0·369	−1·041	0
H_b	1·008	−0·366	0·516	0·896	−0·369	0·520	0·903
H_c	1·008	−0·366	0·516	−0·896	−0·369	0·520	−0·903
C	12·01	0	0	0	0	0	0
O	16·00	1·427	0	0	22·86	0	0
H_d	1·008	1·746	0·903	0	1·760	0·910	0
Sum	32·03				23·51	0·909	0

Atom	$m_i X_i^2$	$m_i Y_i^2$	$m_i Z_i^2$	$m_i X_i Y_i$	$m_i Y_i Z_i$	$m_i X_i Z_i$
H_a	0·135	1·074	0	0·381	0	0
H_b	0·135	0·268	0·809	−0·191	0·466	0·331
H_c	0·135	0·268	0·809	−0·191	−0·466	−0·331
C	0	0	0	0	0	0
O	32·64	0	0	0	0	0
H_d	3·072	0·822	0	1·589	0	0
Sum	36·12	2·432	1·618	1·588	0	0

$$\sum m_i x_i^2 = 36·12 - (23·51^2/32·03) = 18·87$$
$$\sum m_i y_i^2 = 2·432 - (0·909^2/32·03) = 2·406 \quad \text{(by equation 6.2.22)}$$
$$\sum m_i z_i^2 = 1·618 - 0 = 1·618$$

$$\sum m_i x_i y_i = 1·588 - (23·51 \times 0·909/32·03) = 0·921$$
$$\sum m_i y_i z_i = \sum m_i x_i z_i = 0 \quad \text{(by equation 6.2.23)}$$

given in Table 6.4. The moments and products of inertia are then:

$$I_{xx} = 2·406 + 1·618 = 4·026$$
$$I_{yy} = 18·87 + 1·62 = 20·49$$
$$I_{zz} = 18·87 + 2·406 = 21·28$$
$$I_{xy} = 0·921$$
$$I_{yz} = I_{xz} = 0$$

The moment of inertia product is then

$$ABC = \begin{vmatrix} 4{\cdot}026 & -0{\cdot}92 & 0 \\ -0{\cdot}92 & 20{\cdot}49 & 0 \\ 0 & 0 & 21{\cdot}28 \end{vmatrix}$$

$$= 21{\cdot}28(4{\cdot}026 \times 20{\cdot}49 - 0{\cdot}92^2)$$

$$= 1740 \text{ amu}^3 \text{ Å}^6$$

$$= 7{\cdot}96 \times 10^{-138} \text{ kg}^3 \text{ m}^6$$

The partition function for methanol at 300 K is

$$q_{\text{rot(CH}_3\text{OH)}} = 21{\cdot}931 \times 10^{67} \times (7{\cdot}96)^{1/2} \times 10^{-69} \times 3^{3/2} \times 10^3$$

$$= 3{\cdot}22 \times 10^3$$

The symmetry number

In considering the rotational partition function, and the allowed rotational energy levels we have taken no account of molecular symmetry. This has an important influence on the allowable quantum states for rotation of di- and polyatomic molecules and necessitates the introduction of a so-called symmetry number into the formulae for rotational partition functions (6.2.6), (6.2.9) and (6.2.7). The nature of the correction factor may be explained by reference to the hydrogen molecule.

According to quantum theory the two H-nuclei may have spin quantum numbers $+\frac{1}{2}$ or $-\frac{1}{2}$. That is there are two possible spin quantum states for the proton with corresponding nuclear wave functions which we can write as ψ^+ and ψ^-. Four possible nuclear wave functions are then possible for H_2 which can be written, denoting the two H atoms by a and b,

Symmetrical	$\psi^+(a)\psi^+(b)$	$+1$
	$\psi^+(a)\psi^-(b) + \psi^-(a)\psi^+(b)$	0
	$\psi^-(a)\psi^-(b)$	-1
Antisymmetrical	$\psi^+(a)\psi^-(b) - \psi^-(a)\psi^+(b)$	0

The z-component of the nuclear spin is given after each wave function. The wave functions split into two groups. Three are symmetric with regard to exchange of nuclei (i.e. do not change sign when the labels

a and b are interchanged) and one is antisymmetric. All four states have the same energy, or very nearly so.

Now the total wave function for the nuclei of the H_2 molecule must include rotational and vibrational factors, for these types of motion are essentially nuclear not electronic. It is a fundamental property of protons (and other fundamental particles) that the total wave function for exchange of protons must be antisymmetric. That is $\psi_{spin} \times \psi_{rot} \times \psi_{vib}$ must be antisymmetric with respect to exchange of the labels a and b. Now the vast majority of H_2 molecules at reasonable temperatures are in the ground vibrational state which has a symmetric wave function. Thus to obtain overall antisymmetry the symmetrical spin wave functions must be combined with antisymmetric rotational wave functions ($j = 1, 3, 5, \ldots$), while the antisymmetric nuclear wave function must be combined with symmetric rotational wave functions ($j = 0, 2, 4, \ldots$).

The states with symmetric spin wave functions are called ortho states, and those with antisymmetric spin wave functions are called para states. The distinction between the ortho and para states has nothing to do with whether the spins of the two nuclei are the same or are opposed. The spin degeneracy of the ortho states is therefore three, and of the para states unity.

The two types of H_2 molecule must clearly have different nuclear and rotational partition functions namely

$$q_{rot(o\text{-}H_2)} = \sum_{j=1,3..} (2j + 1) \exp\left[-j(j + 1)\theta_r/T\right]; \qquad q_{nuc(o\text{-}H_2)} = 3$$

$$q_{rot(p\,H_2)} = \sum_{j=0,2..} (2j + 1) \exp\left[-j(j + 1)\theta_r/T\right]; \qquad q_{nuc(p\text{-}H_2)} = 1$$

$$(6.2.24)$$

For hydrogen θ_r is 87·5 K, and the two rotational partition functions differ markedly at low temperatures. Accordingly the thermodynamic properties of ortho- and para-hydrogen differ. The two forms interconvert very slowly in the absence of a catalyst, but quite rapidly in the presence of activated charcoal. By cooling normal hydrogen over activated charcoal in liquid nitrogen it is possible to produce nearly pure para-hydrogen, the form which has the rotational state of lowest energy. At very high temperatures the two partition functions converge, and because of the differences in spin degeneracy the high temperature equilibrium mixtures contain a 3:1 ratio of ortho- to para-forms. By measuring the thermodynamic properties of this high temperature mixture, and of pure para-H_2, the properties of both pure forms may be found. The agreement between theory and experiment (see Chapter 9) is one of the more elegant justifications of the postulates of molecular thermodynamics.

At high temperatures, when θ_r/T small, the distinction between the partition functions for ortho- and para-forms of hydrogen, and of any other symmetrical diatomic molecule disappear, and we obtain

$$q_{rot(o\text{-}H_2)} = q_{rot(p\text{-}H_2)} = q_{rot(H_2)}$$

$$= \frac{1}{2} \sum_{\text{all } j} \{(2j + 1) \exp [-j(j + 1) \theta_r/T]\}$$

$$= \frac{8\pi^2 IkT}{2h^2} \qquad (6.2.25)$$

The factor, one-half, which is missing from equation (6.2.9), is seen to be a necessary part of the rotational partition function. It arises from the symmetry of the molecule which demands that only every other rotational quantum state is accessible to any one nuclear isomer. All symmetrical diatomic molecules possess symmetric and antisymmetric spin wave functions which define ortho- and para-states, but only for hydrogen is θ_r sufficiently high that the ortho- and para-forms have distinct properties. For all other symmetrical diatomic molecules the properties of the two forms are indistinguishable. The factor 2 by which q_{rot} must be divided for symmetrical diatomic molecules is called the symmetry number. For unsymmetrical diatomic molecules all rotational states are accessible to every molecule, and the symmetry number is unity.

Similar but more complex considerations apply to polyatomic molecules. The spin wave functions, and rotational/vibrational wave functions may be divided into various symmetry classes which restrict the rotational wave functions which can be allocated to any given molecule because the total nuclear wave function must be antisymmetric with respect to exchange of protons and neutrons. The problem is discussed in somewhat greater detail for methane by Fowler and Guggenheim. The final result of these considerations is that a symmetry number arises which may be defined as follows:

The symmetry number, denoted by σ, is the number of configurations of a molecule which can be generated from any starting configuration by rotation(s) of a given type, which are distinguishable when atoms of the same type are labelled, but are indistinguishable when the labels are removed.

The origin of the symmetry number is formally not unlike that of the factor $N!$ which is introduced into the formula for the partition function of an assembly of non-localized systems, for $N!$ is simply the number of complexions of the assembly which are distinguishable when systems of the same type are labelled, but which are indistinguishable when the labels are removed.

The formulae which have previously been given for the rotational partition functions must now be corrected by inclusion of the symmetry number.

$$q_{rot(1\text{-}D)} = \frac{\sqrt{\pi}}{\sigma}\left(\frac{8\pi^2 IkT}{h^2}\right)^{1/2} \tag{6.2.26}$$

$$q_{rot(2\text{-}D)} = \frac{1}{\sigma}\frac{8\pi^2 IkT}{h^2} \tag{6.2.27}$$

$$q_{rot(3\text{-}D)} = \frac{\sqrt{\pi}}{\sigma}\left(\frac{8\pi^2 kT}{h^2}\right)^{3/2}(ABC)^{1/2} \tag{6.2.28}$$

The method of evaluating the symmetry number is best illustrated by examples.

Example 6.5 The symmetry number for internal rotation of dimethyl acetylene.

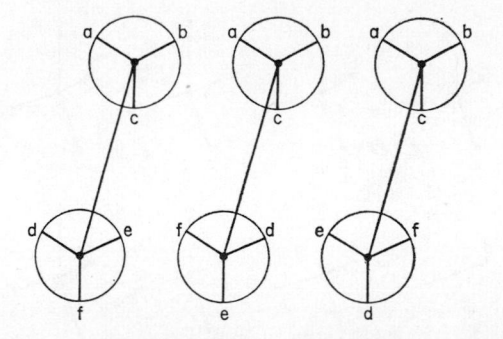

Figure 6.11 Configurations obtainable by internal rotation of methyl groups in dimethyl acetylene showing the internal rotational symmetry number to be 3.

We are concerned here with the rotation of one methyl group against the other which may for the purpose of the argument by regarded as fixed. Figure 6.11 shows the three configurations, including the starting one, which can be generated by rotation of one methyl group against the other. They differ when the H atoms are labelled, but become indistinguishable when the labels are removed. The symmetry number for *internal rotation* is thus three.

$$\sigma_{f.i.r} = 3$$

The correct partition function for internal rotation of dimethyl acetylene at 300 K is therefore

$$q_{rot(1-D)} = 7{\cdot}90/3 = 2{\cdot}63$$

Example 6.6   The symmetry numbers for ethylene and methane.

Ethylene cannot undergo internal rotation under normal conditions since the barrier to twisting the molecule is about 179 kJ mol^{-1}. We are therefore concerned only with overall rotations of the molecule. Figure 6.12 shows that there are four configurations (counting the original)

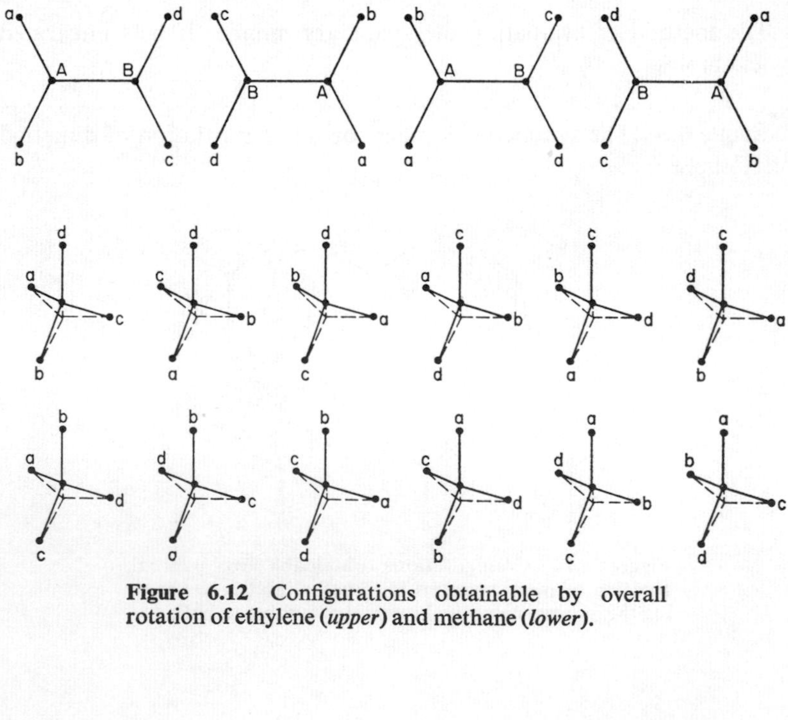

Figure 6.12 Configurations obtainable by overall rotation of ethylene (*upper*) and methane (*lower*).

which can arise from any starting configuration, which differ when the H and C atoms are labelled but become indistinguishable when the labels are removed. Thus $\sigma_{C_2H_4} = 4$.

For methane there are twelve such configurations; $\sigma_{CH_4} = 12$. The correct partition functions for overall rotation of ethylene and methane are thus

$$q_{rot(C_2H_4)} = 2 \cdot 67 \times 10^3/4 = 670$$

$$q_{rot(CH_4)} = 44 \cdot 2/12 = 4 \cdot 7$$

For methanol in the absence of internal rotation (barrier height $4 \cdot 5$ kJ mol^{-1}) no additional configurations can be generated from any starting configuration which are distinguishable from the original when like atoms are labelled but indistinguishable in the absence of labels. The symmetry number is thus unity, and the partition function given in Example 6.4 is correct.

Example 6.7 The complete symmetry number of dimethyl acetylene.

Independent symmetry numbers arise from the internal and overall rotations of dimethyl acetylene. In considering the overall rotational symmetry number the two methyl groups are taken as fixed relative to each other. There are six distinguishable orientations when like atoms are labelled which become indistinguishable when the labels are removed. The first three are obtained by rotating the molecule about an axis through the four C atoms. The second three are obtained by a similar rotation after turning the molecule end over end. Since each of these six orientations may be combined with each of the three internal orientations the total symmetry number is obtained by multiplication:

Symmetry number for internal rotation $= 3$

Symmetry number for overall rotation $= 6$

Complete symmetry number $= 3 \times 6 = 18$

The complete rotational partition function for dimethyl acetylene is finally obtained by multiplying the partition function for internal rotation by that for overall rotation. The principal axes of rotation for dimethyl acetylene can be seen by inspection (Figure 6.4). The bond lengths are $r(C—H) = 1 \cdot 095$, $r(C—C) = 1 \cdot 47$ and $r(C\equiv C) = 1 \cdot 20$ Å. The angles in the methyl group are 109·5°. The moments of inertia about the principal axes of rotation are then

$$A = 6·420 \text{ amu Å}^2$$

$$B = C = 150·4 \text{ amu Å}^2$$

The moment of inertia for internal rotation is

$$I = 1·604 \text{ amu Å}^2$$

The complete partition function for all rotations is

$$q_{\text{rot(4-D)}} = \frac{\pi}{\sigma} \left(\frac{8\pi^2 kT}{h^2}\right)^2 (IABC)^{1/2}$$

$$= \frac{\pi}{18} \left(\frac{8\pi^2 kT}{h^2}\right)^2 \times \frac{(1·604 \times 6·420 \times 150·4^2)^{1/2} \times 10^{-40}}{6·022^2 \times 10^{52}}$$

$$= 1·28 \times 10^4$$

6.3 THE VIBRATIONAL PARTITION FUNCTION

The potential function and the energy levels for the quantum harmonic oscillator are shown in Figure 6.13. The energy levels follow equation (6.3.1)

$$\epsilon_n = (n + \tfrac{1}{2}) h\nu = (n + \tfrac{1}{2}) h\omega c \qquad (6.3.1)$$

ν is the fundamental frequency of vibration, that is, the number of vibrational cycles per unit time; ω is the frequency in wave numbers, that is, the number of cycles per unit length; and c is the velocity of light. As with the classical harmonic oscillator the frequency remains constant independent of the energy but the amplitude of the oscillation increases. In wave mechanical terms the extension of the wave function increases with the energy, as does the number of nodes in the wave function.

Because of the uncertainty principle the quantum harmonic oscillator cannot have a total energy of zero, for then both the position and the momentum of the vibrating particle or system would be precisely fixed. As a result the lowest quantum state, called the *ground vibrational state*, has an energy of $h\nu/2$ above the classical energy zero. In many applications it is convenient to measure the energies of the various quantum

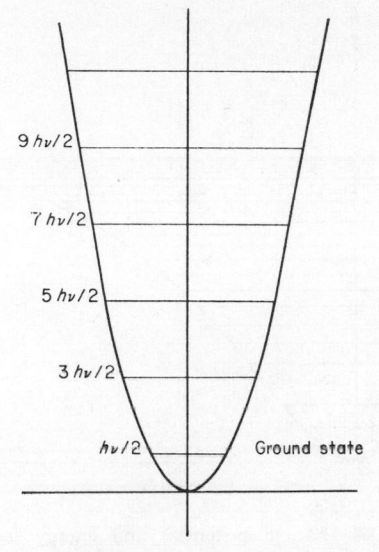

$9\,h\nu/2$

$7\,h\nu/2$

$5\,h\nu/2$

$3\,h\nu/2$

$h\nu/2$ Ground state

Figure 6.13 Potential function and energy levels of quantized harmonic oscillator.

states from the ground state rather than from the classical energy zero. Equation (6.3.1) is then modified to

$$\epsilon'_n = nh\nu = nh\omega c \qquad (6.3.2)$$

For any real molecule, say a diatomic molecule, the vibration, whether considered in classical or quantum terms, cannot be truly harmonic as this implies complete symmetry of the potential function with respect to extension or contraction. For small displacements from equilibrium, molecules behave as if they were harmonic oscillators, but clearly a bond can be indefinitely extended, but cannot be indefinitely compressed.

Figure 6.14 shows the Morse potential function which gives a more realistic representation of the potential for a diatomic molecule. It has the form

$$U(r) = D_e(1 - \exp[-b(r - r_0)])^2 \qquad (6.3.3)$$

where r is the internuclear distance, and r_0 is the distance for minimum potential energy, D_e is the dissociation energy measured from the classical energy zero, that is the bottom of the potential well, and b is a

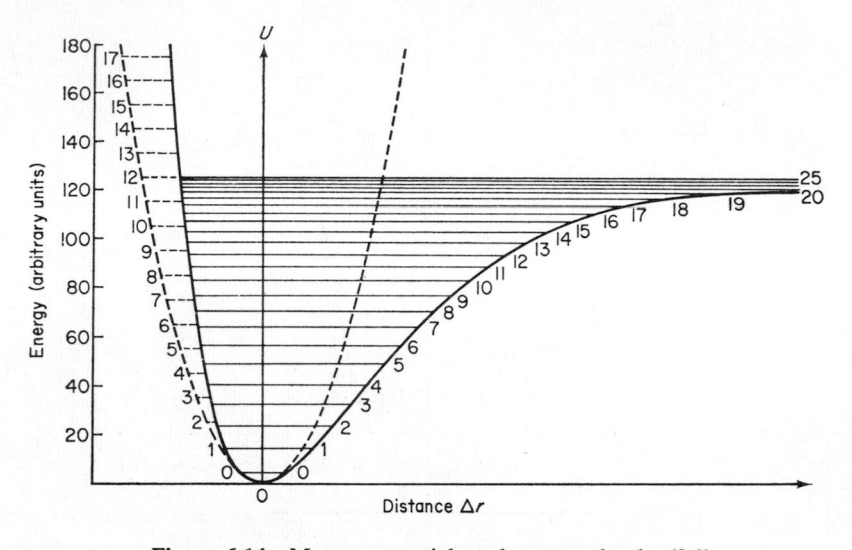

Figure 6.14 Morse potential and energy levels (full lines). Harmonic potential and energy levels (broken lines). Parameters for Morse potential: $D_e = 125$, $h\nu_e = 10$, $x_e = 0.02$.

constant. This function has the advantage that it allows explicit solutions to the Schrödinger equation. The allowed energy levels are given by (6.3.4).

$$\epsilon_n = h\nu_e \{(n + \tfrac{1}{2}) - x_e(n + \tfrac{1}{2})^2\} \qquad (6.3.4)$$

where the parameters ν_e and x_e are expressed in terms of b and D_e as

$$\nu_e = \frac{b}{\pi} \left(\frac{D_e}{2\mu}\right)^{1/2}, \qquad x_e = \frac{h\nu_e}{4D_e} \qquad (6.3.5)$$

where μ is the reduced mass of the oscillator. x_e is called the anharmonicity constant. It is readily seen from equation (6.3.4) and Figure 6.14 that at low energies the Morse potential approximates to the harmonic oscillator potential, but that as the energy increases it diverges more and more so that extension of the bond by a given amount is progressively easier than compression by the same amount. Equation (6.3.4) and Figure 6.14 also show that the energy levels of the Morse oscillator become closer as n increases, and that for sufficiently large values of n ϵ_n should start decreasing again. It is readily shown that ϵ_n is a maximum when $\epsilon_n = D_e$, and $n = 1/2x$. Quantum states with n in excess of $1/2x$ cannot of course correspond to reality. Indeed at values of n close to $1/2x$, the Morse curve gives an increasingly poor representation of the real situation. For Figure 6.14 $x_e = 0.02$. Most real molecules have x_e between 0.002 and 0.02.

The partition function for the quantum harmonic oscillator when the energies are measured from the classical zero is

$$q_{\text{vib}} = \sum_{n=0}^{\infty} \exp\left[-(n + \tfrac{1}{2})\, h\nu/kT\right] = \sum \exp\left[-(n + \tfrac{1}{2})\, \theta_v/T\right]$$

$$(6.3.6)$$

where $\theta_v = h\nu/k = h\omega c/k$ is the "characteristic vibrational temperature" of the oscillator. Values of ω and θ_v for some diatomic and linear polyatomic molecules are given in Table 6.2. θ_v/T is a measure of the spacing of the energy levels compared to kT. As shown in Chapter 4 at equation (4.2.21) the terms in the summation of (6.3.6) are in geometrical progression with a common ratio less than unity. The sum is therefore

$$q_{\text{vib}} = \frac{\exp\left[-h\nu/2kT\right]}{1 - \exp\left[-h\nu/kT\right]}$$

$$= \frac{\exp\left[-\theta_v/2T\right]}{1 - \exp\left[-\theta_v/T\right]} \qquad (6.3.7)$$

If the energies of the states are measured from the ground state, equation (6.3.2) is used and leads to

$$q'_{\text{vib}} = \sum_{n=0}^{\infty} \exp\left[-(\theta_v/T)n\right] = (1 - \exp\left[-\theta_v/T\right])^{-1}$$

$$= (1 - \exp\left[-h\nu/kT\right])^{-1}$$

$$(6.3.8)$$

The partition function for the Morse oscillator cannot be evaluated explicitly, but the summation may be carried out term by term using a computer. The series must be truncated at the value for n for which ϵ_n is a maximum; that is when n is the last integer before $1/2x$. The partition function is thus

$$q_{\text{vib(Morse)}} = \sum_{n=0}^{1/2x} \exp\left[-h\nu_e\{(n + \tfrac{1}{2}) - x_e(n + \tfrac{1}{2})^2\}/kT\right] \quad (6.3.9)$$

Since the Morse potential is probably accurate only at energies somewhat below D_e this partition function will be reliable provided that the series converges well before the final term. This is so when kT is less than about $0 \cdot 125 D_e$. The difference between the two partition functions is best evaluated if the energies of the quantum states are measured from the ground state. Making this alteration the energies for the Morse oscillator are

$$\epsilon'_n = h\nu_e\{n - n(n + 1)x_e\} \qquad (6.3.10)$$

If x_e is replaced by the value given in equation (6.3.5) we obtain

$$\epsilon'_n = h\nu_e \left\{n - n(n + 1)\frac{h\nu_e}{4D_e}\right\} \qquad (6.3.11)$$

The partition function then becomes

$$q'_{\text{vib(Morse)}} = \sum_{n=0}^{1/2x} \exp\left[-\left(\frac{\theta_v}{T}\right)n + \left(\frac{\theta_v}{T}\right)^2\left(\frac{kT}{4D_e}\right)n(n + 1)\right]$$
$$(6.3.12)$$

This is then to be compared with the expression for the Harmonic oscillator (6.3.8).

For both series the leading term is unity, and at the limit of zero temperature both partition functions are unity. Provided that the summation of (6.3.12) converges well before the limit $n = 1/2x$, the difference between $q'_{\text{vib (Morse)}}$ and $q'_{\text{vib (harmonic)}}$ must be largely a function of kT/D_e. This is illustrated in Figure 6.15 which shows the fractional error which results from the use of the harmonic approximation instead of the somewhat more accurate Morse approximation. The error for $kT = 0 \cdot 1 D_e$, is seen to be about 5%. Since D_e is typically $100 - 400$ kJ mol^{-1} for single bonds, the temperature at which $kT = 0 \cdot 1 D_e$ will be in the range $10/R$ to $40/R$ where $R = 0 \cdot 0083$ kJ mol^{-1}, that is between 1250 and 5000 K. Such temperatures are sufficiently high that the harmonic oscillator approximation may be considered accurate for nearly all chemical applications of molecular thermodynamics.

In evaluating the translational and rotational partition functions we were able to replace the exact summations by integrations without introduction of significant error. Although this is not generally possible

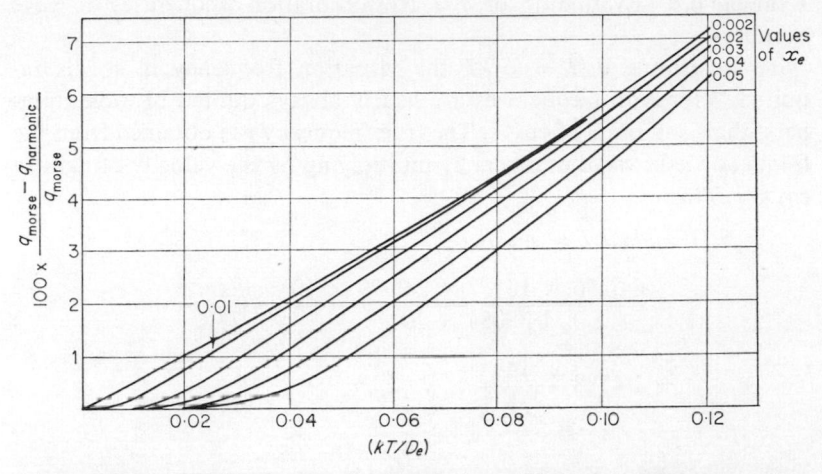

Figure 6.15 Fractional error in harmonic oscillator approximation when compared to Morse approximation for various values of x_e given on lines. The curves terminate at values of (kT/D_e) where the convergence of the Morse partition function becomes poor.

for vibrational degrees of freedom, it is nevertheless worth evaluating the integrated form.

$$q_{\text{vib}} = \int_0^\infty \exp\left[-(n + \tfrac{1}{2})(\theta_v/T)\right] dn = (T/\theta_v) \exp\left[-\theta_v/2T\right]$$

$$(6.3.13)$$

The approximation is in error by 1% when $\theta_v/T = 0.02$. Since the exponential term in (6.3.13) is then 1·01, the result becomes:

$$q_{\text{vib}} = T/\theta_v = kT/h\nu \qquad (6.3.14)$$

Typically vibration frequencies in molecules fall within the range 100 to 3000 cm^{-1} (the unit cm^{-1} is commonly called the "wave number", and in the context of vibration frequency implies the number of vibrations per cm). It is shown in Example 6.8 that θ_v/T for any frequency ω is obtained from the formula

$$\frac{\theta_v}{T} = \frac{h\nu}{kT} = 1.4388 \frac{\omega/\text{cm}^{-1}}{T/\text{K}} \qquad (6.3.15)$$

Thus for typical molecular vibrations at 300 K, θ_v/T falls in the range 0·5 to 15. Even at 1000 K, θ_v/T will be between 0·15 and 6. Under such conditions the approximation (6.3.14) is never sufficiently accurate to be useful.

Example 6.8 Evaluation of θ_v/T from vibration frequencies in wave numbers.

To determine $\theta_v/T = h\nu/kT$ the vibration frequency in s^{-1} is required. Vibration frequencies are nearly always quoted in wave numbers, that is in units of cm^{-1}. The true frequency ν is obtained from the frequency ω in wave numbers by multiplying by the velocity of light in $cm\ s^{-1}$. Thus

$$\theta_v/T = h\nu/kT = h\omega c/kT$$

$$= \frac{6\cdot6256 \times 10^{-34} \times 2\cdot9979 \times 10^{10}}{1\cdot38054 \times 10^{-23}} \frac{\omega/cm^{-1}}{T/K}$$

$$= 1\cdot4388 \frac{\omega/cm^{-1}}{T/K}$$

The vibrations of polyatomic molecules are complex but may, to a good approximation, like other types of motion, be resolved into independent components. The allowable vibrations are called the normal vibrations or normal vibrational modes of the molecule. Each normal vibration involves the oscillation of all atoms of the molecules about their mean positions with a common frequency unless symmetry forbids this. The method used to determine the magnitudes and directions of these coupled vibrations of the atoms are covered in books on dynamics and elsewhere. For the present purpose it is sufficient to recognize that such normal vibrations occur, that to a first approximation they are independent and harmonic, and that their number is strictly defined.

For a molecule containing n atoms, $3n$ positional coordinates are required to specify the positions of all the atoms of the molecule. Since any molecule by definition may be regarded as a relatively rigid and stable framework of atoms with a well-defined shape and centre of mass, the $3n$ coordinates fall naturally into three nearly independent groups: coordinates of the centre of mass which give the general position of the molecule in space, angular coordinates which give its orientation in space, and internal coordinates relative to axes fixed within the molecule which describe the internal configuration. Changes in the first group of coordinates reflect translational motion of the molecule, changes in the second, rotational motion, and changes in the third, vibrational or internal rotational motion.

All molecules require 3 cartesian coordinates to specify the position of the centre of mass. Linear molecules require two angles to specify their orientation in space, while non-linear molecules require three such angles. Thus linear molecules have $3n - 5$ coordinates which may be used to specify the internal state, while non-linear molecules have $3n - 6$. That is there are $3n - 5$ vibrational modes for linear molecules (internal rotations are impossible) and $3n - 6$ vibrational + internal rotational modes in non-linear molecules.

The normal modes of vibration in the water and carbon dioxide are illustrated in Figure 6.16. Both molecules have three atoms, but water

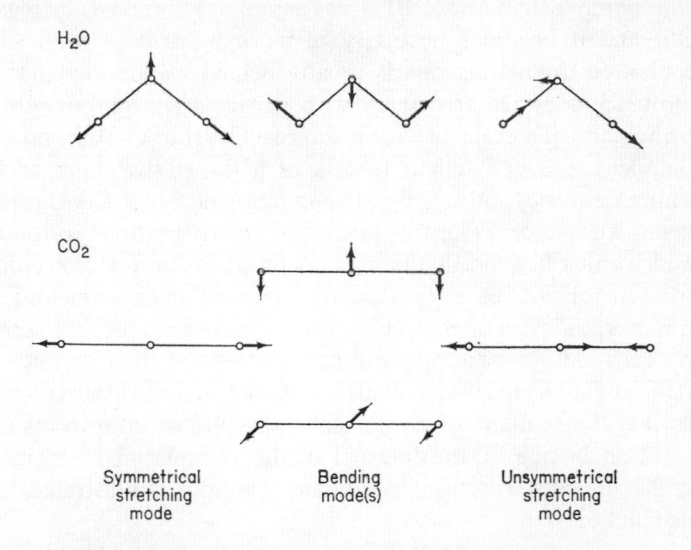

| Symmetrical stretching mode | Bending mode(s) | Unsymmetrical stretching mode |

Figure 6.16 Normal vibrational modes of water and carbon dioxide molecules.

being bent has three modes while carbon dioxide being linear has four. Both molecules possess a symmetrical and unsymmetrical stretching mode. The third type of vibration is bending. The water molecule has a unique bending mode, but carbon dioxide can clearly bend either in the plane of the page or perpendicular to it. This mode is therefore doubly degenerate. It is interesting to observe that the analog in water of the degenerate bending mode of carbon dioxide, is a combination of the unique bending mode and rotation about the x-axis. The correlation is emphasized by noting that if the carbon dioxide molecule had equal amounts of energy in the two bending modes, and if the vibrations were suitably phased, the molecule would appear to be bent and to be rotating about the x-axis with a period equal to its vibration frequency. The initially clear distinction between the linear and bent triatomic molecules is thus to some extent blurred.

The frequencies of the normal modes of simple molecules (say up to about 10 atoms) can often be found from their infra-red and Raman spectra, but the problem is complex and the assignment of frequencies cannot always be made with certainty (see for example Herzberg, *Infra-red and Raman Spectra of Polyatomic Molecules*). For molecules containing more than about 10 atoms assignment becomes increasingly difficult and it becomes necessary to work by analogy with simpler molecules. In doing this one is greatly helped by the fact that most vibrational modes are associated with particularly vigorous vibration of certain parts of a molecule, for instance a C—H or C—C bond. When this happens it is convenient to talk of a C—H stretching or C—C stretching frequency, of a CH_2 wagging frequency, or a C=C torsional frequency and so on. The frequencies of such vibrations in unknown molecules can often be deduced from those in known molecules. In the great majority of cases they are the same within experimental error.

An important class of molecules, not so far mentioned, are activated complexes which possess one internal mode fewer than normal molecules of similar shape. Activated complexes or transition state complexes may be thought of as normal molecules in all respects except that motion in one of the internal modes is replaced by translation along the so-called reaction coordinate. The idea is illustrated by the simple reaction

$$Cl^* + H\text{—}Cl \longrightarrow Cl^*\text{--}H\text{--}Cl \longrightarrow Cl^*\text{—}H + Cl$$

The molecule Cl^*--H--Cl is the activated complex. It may be described as half way between the reactants and products. The complex is linear and if a normal molecule would have 4 vibrational modes, namely a symmetrical stretch, an unsymmetrical stretch and a doubly degenerate bending mode. In fact the unsymmetrical stretching mode represents the progress of the reaction for it implies the coupled extension of the H--Cl bond and contraction of the Cl^*--H bond. Linear activated complexes therefore have $3n - 6$ vibrational modes, while non-linear complexes have $3n - 7$ internal modes of motion (vibrations and internal rotations).

In the harmonic approximation the normal vibrational modes of any molecule are orthogonal, that is independent, and the complete vibrational partition function, by the multiplication theorem, is the product of the partition functions for the normal modes.

For a non-linear molecule with no internal rotations the complete vibrational partition function is

$$q_{\text{vib}} = \prod_{i=1}^{3n-6} \left\{ \frac{\exp\left[-\theta_{v,\,i}/2T\right]}{1 - \exp\left[-\theta_{v,\,i}/T\right]} \right\} \qquad (6.3.16)$$

If the energies of the vibrational states are measured from the ground vibrational states of the modes then

$$q'_{\text{vib}} = \prod_{i=1}^{3n-6} \left\{ 1 - \exp\left[-\frac{\theta_{v,\,i}}{T}\right] \right\}^{-1} \qquad (6.3.17)$$

where $\theta_{v,\,i} = h\nu_i/k$, ν_i being the fundamental frequency of the i'th mode. In most applications it is convenient to use equation (6.3.17) rather than (6.3.16).

As with diatomic molecules, the harmonic oscillator is only an approximation to the true vibrational motion of any mode of a complex molecule. To a greater or lesser extent all modes are anharmonic, and where they predominantly involve the stretching of a single bond, the Morse equation will give a reasonable representation. Nevertheless as was shown above, little error is introduced in most chemical applications if vibrations are assumed to be harmonic. It is only when dealing with highly excited molecules possessing large amounts of internal vibrational energy that corrections for anharmonicity become necessary. Situations where this may occur arise in association reactions of radicals where up to 400 kJ mol^{-1} may be released as internal vibrational energy. Yet even this rather large amount of energy when distributed evenly throughout a large molecule rarely causes much excitation of any particular mode of internal motion.

Example 6.9 The vibrational partition function for carbon dioxide at 300 K and 1000 K.

The vibrational frequencies for carbon dioxide are

symmetrical stretch $\quad \omega_1 = 1351 \text{ cm}^{-1}$

bend (degeneracy = 2) $\quad \omega_2 = 672\cdot2 \text{ cm}^{-1}$

unsymmetrical stretch $\quad \omega_3 = 2396 \text{ cm}^{-1}$

The partition function for vibration is then, by equation (6.3.16),

$$q_{\text{vib}} = q_{\text{vib(1)}} \times q_{\text{vib(2)}}^2 \times q_{\text{vib(3)}}$$

$q_{\text{vib(2)}}$ appears squared because of the degeneracy of this mode. We evaluate $q_{\text{vib(2)}}$ for 1000 K to illustrate the calculation. The other values of the component q_{vib}'s are listed in Table 6.5. Energies are measured from the ground states of each mode, thus the partition function for any mode is given by equation (6.3.8)

$$q'_{\text{vib(2)}} = (1 - \exp[-h\omega_2 c/kT])^{-1}$$

at 1000 K

$$h\omega_2 c/kT = 1\cdot4388 \times 672\cdot2/1000$$

$$= 0\cdot9672$$

$$q'_{\text{vib(2)}} = (1 - \exp[-0\cdot9672])^{-1}$$

$$= 1\cdot6133$$

The complete vibrational partition function for the two temperatures is obtained from the data of the table as

$$q'_{\text{vib}}(300 \text{ K}) = 1\cdot00154 \times 1\cdot04145^2 \times 1\cdot00001$$

$$= 1\cdot0863$$

$$q'_{\text{vib}}(1000 \text{ K}) = 1\cdot1671 \times 1\cdot6133^2 \times 1\cdot0329$$

$$= 3\cdot1376$$

Table 6.5   Vibrational partition function for carbon dioxide

Frequency ω/cm^{-1}	$h\omega c/kT$		q_{vib}	
	300 K	1000 K	300 K	1000 K
1351 [1][a]	6·480	1·944	1·00154	1·1671
672·2 [2]	3·224	0·9672	1·04145	1·6133
2396 [1]	11·491	3·4472	1·00001	1·0329

[a] Figures in [ ] are the degeneracies of the modes.

6.4 THE ELECTRONIC PARTITION FUNCTION AND CHANGE OF ENERGY ZERO

Only for the simplest atoms can explicit expressions be written for the electronic energy levels. It is thus generally not possible to evaluate an electronic partition in closed form. Further molecular shapes often change when they become electronically excited and separation of the electronic and other internal energy levels is not possible. The concept of an electronic partition function thus has limited usefulness, for the multiplication theorem cannot generally be applied.

If a molecule has low lying electronic energy states the complete partition function has the form

$$q = q_{trans} \left\{ g_0 q_{0(v,r)} \exp\left[-\frac{\epsilon_{0(el)}}{kT}\right] + g_1 q_{1(v,r)} \exp\left[-\frac{\epsilon_{1(el)}}{kT}\right] + \ldots \right\}$$

(6.4.1)

where g_0, g_1 etc. are the degeneracies of the electronic energy levels, $q_{0(v,r)}$ etc. are the vibrational–rotational partition functions for each electronic state, and $\epsilon_{0(el)}$ etc. are the energies of the electronic energy levels.

Equation (6.4.1) may be written in the simplified form

$$q = a_0 \exp\left[-\epsilon_0/kT\right] + a_1 \exp\left[-\epsilon_1/kT\right] + \ldots$$ (6.4.2)

where the a_i's take care of the electronic degeneracies, and all other contributions to the molecular partition functions, and where the subscript "el" has been dropped from the electronic energies.

Although ϵ_0 has been written as the energy of the lowest state, there is no way of assigning an absolute value to it. Indeed the only information which we have about the electronic energy levels are the differences between them, and thus values of quantities of the type

$$\Delta\epsilon_i = \epsilon_i - \epsilon_0$$ (6.4.3)

Writing equation (6.4.2) in terms of ϵ_0 and the $\Delta\epsilon_i$ gives

$$q = \exp\left[-\epsilon_0/kT\right] \{a_0 + a_1 \exp\left[-\Delta\epsilon_1/kT\right] + \ldots\}$$

$$= q' \times \exp\left[-\epsilon_0/kT\right]$$ (6.4.4)

where q' is the partition function evaluated with energies which are measured from the ground state of the molecule as shown in Figure

6.17. q' is called the "conventional partition function" to distinguish it from a partition function evaluated with energies measured from some arbitrary zero.

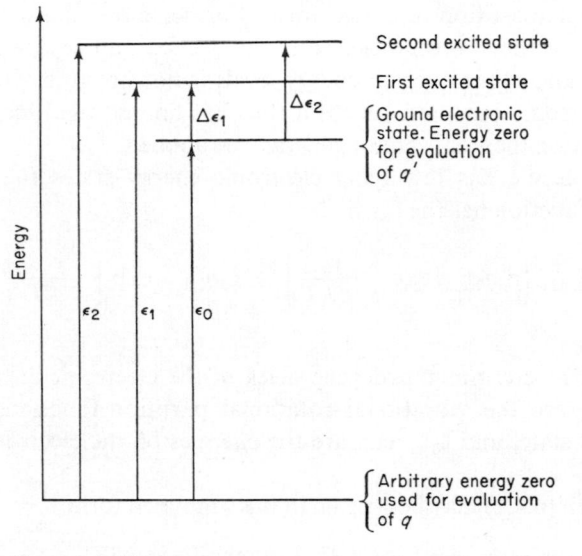

Figure 6.17 Energies above ground state, $\Delta\epsilon$, and energies above some arbitrary zero, ϵ.

Equation (6.4.4) may be given a more general interpretation for it relates partition functions for the same species evaluated with energies based upon different zero points:

q (evaluated with energies based upon an arbitrary zero)

$$= \exp\left[-\Delta\,\epsilon_0^\circ/kT\right] \times q \text{ (evaluated with energies based upon an energy zero higher by } \Delta\epsilon_0^\circ) \qquad (6.4.5)$$

PROBLEMS

6.1 Calculate the translational partition function for CH_4 at 112 K, the boiling point, and at 1000 K in a volume of 1 litre (10^{-3} m^3).

6.2 Calculate the translational partition function for a two-dimensional ideal gas film of benzene, 0·01 m² in area on say a mercury surface at 300 K.

6.3 Calculate the symmetry numbers of the following molecules: ethane (no free internal rotation), 2-methyl propane (no free internal rotation), benzene, benzoquinone, toluene, nickel carbonyl, sulphur hexafluoride, dimethyl cadmium (assume free internal rotation, $C-Cd-C$ angle $= 180°$).

6.4 Calculate the rotational partition functions for the CN, OH and ClO radicals at 300 K given the bond lengths $r(C-N) = 1·157$ Å, $r(O-H) = 0·971$ Å, $r(Cl-O) = 1·49$Å.

6.5 Calculate the rotational partition functions of SF_6 and BF_3 at 300 K, given the bond lengths $r(S-F) = 1·58$ Å, $r(B-F) = 1·295$ Å.

6.6 Show that for small displacements from equilibrium the Morse potential function, equation (6.3.3), reduces to the harmonic potential function of the form $U = \frac{1}{2}f(r-r_0)^2$. Hence taking the frequency as $\nu = (1/2\pi)(f/\mu)^{1/2}$ prove equation (6.3.5) for ν_e.

6.7 Show, using equation (6.3.4), that ϵ_n is a maximum for $n + \frac{1}{2} = 1/2x_e$.

6.8 How many vibrational modes does the acetylene molecule possess? Draw the fundamental modes qualitatively and indicate which will be degenerate.

6.9 The H_2O_2 molecule is non-linear. Assuming it is planar and staggered, draw the fundamental modes qualitatively. Which if any are degenerate? How many modes should there be?

6.10 H_2O_2 dissociates unimolecularly into two OH radicals via an activated complex which may be written as H—O–––––O—H. Which of the modes of internal motion described in Problem 6.9 becomes the reaction mode? What are the remaining modes of vibration of the complex?

6.11 The fundamental vibration frequencies of CCl_4 in cm^{-1} are: 461·5, 217·9 (2), 776 (3), 314·0 (3). The degeneracies are given in brackets. Calculate the vibrational partition functions at 300 K for each mode, and hence obtain the complete vibrational partition function.

The C—Cl bond distance is 1·766 Å. Calculate the rotational partition function as was done for methane in worked Example 6.9. Why broadly speaking is q_{rot} much larger than q_{vib}?

6.12 Equations (6.3.7) and (6.3.8) give alternative equations for the vibrational partition function. Show that they are related by equation (6.4.5).

6.13 Plot against quantum number the probability that a rotator taken at random be found in a rotational energy level of quantum number j, (a) for a one-dimensional rotator and (b) for a two-dimensional rotator. Assume $\theta_r/T = 0·05$ (see equations 4.2.20, 5.0.1 and 5.0.2). Why are the two graphs so different?

6.14 Carry out the same calculation as in Problem 6.13 for a one-dimensional harmonic oscillator for which $\theta_v/T = 0.05$. What would be the result for a two-dimensional oscillator?

(*Note*: to solve the second part find the degeneracy of each energy level by counting the number of ways of distributing n quanta between two oscillators.)

6.15 Write down the equation for the probability of finding a two-dimensional rotator in an energy level of quantum number j. Differentiate $p(\text{energy}, \epsilon_j)$ with respect to j, and so obtain an equation for the most highly populated energy level. Check this equation by applying it to Problem 6.13.

6.16 What are the relative populations of (*a*) the first three rotational energy levels of the HCl molecule at 300 K (for θ_r see Table 6.2), (*b*) the first three vibrational states given the fundamental frequency for HCl to be 2990 cm^{-1}.

At what temperature would 10% of HCl molecules be in the first excited vibrational state.

6.17 At room temperature HCl is predominantly in the ground vibrational state, but as shown by Problem 6.16, several rotational states have significant populations. HCl absorbs infra-red radiation which excites molecules from the ground vibrational state to the first excited state ($n = 0$ to $n = 1$), but concurrently the rotational quantum number must change by $+$ or -1. Derive an expression for the energies of all possible transitions for the various values of j. Assuming that the probability of a transition is proportional to the population of the state being excited, predict the intensity of the lines in the HCl infra-red spectrum. Compare your prediction with a typical spectrum for HCl given in a standard text of physical chemistry, for example Barrow's *Physical Chemistry*, page 264.

CHAPTER 7

CLASSICAL MOLECULAR PARTITION FUNCTIONS

The classical molecular partition function for any type of motion is obtained by defining the accessible regions of phase space, and assigning to each point its correct total energy (that is kinetic + potential energy). If n degrees of freedom are required to specify the geometry of the system then the classical partition function is

$$q = h^{-n} \int \ldots \int dp_1 \, dq_1 \, dp_2 \, dq_2 \ldots dp_n \, dq_n \exp\left[-\epsilon/kT\right]$$

(7.0.1)

where ϵ is the total energy and in general is a function of all the p's and q's (momenta and positions). The factor h^{-n} is introduced in order to satisfy the correspondence principle that quantum and classical theories must converge for large systems, or assemblies of systems.

7.1 THE CLASSICAL TRANSLATIONAL PARTITION FUNCTION FOR A STRUCTURELESS PARTICLE

The energy of a particle translating in a field free box is entirely kinetic and given by

$$\epsilon = (1/2m)\,(p_x^2 + p_y^2 + p_z^2)$$

(7.1.1)

ϵ is independent of the values of q_x, q_y and q_z, except in so far as it becomes infinite for q's representing positions outside the box. This condition is dealt with by limiting the confines of the configurational phase space rather than by introducing a potential energy term. The partition function is then, applying (7.0.1),

$$q = h^{-3} \int\int\int\int\int\int \exp\left[-\frac{p_x^2 + p_y^2 + p_z^2}{2mkT}\right] dp_x \, dp_y \, dp_z \, dq_x \, dq_y \, dq_z$$

137

For a rectangular box the limits of the q's are $0 < q_x < a$, $0 < q_y < b$, and $0 < q_z < c$. The limits of the p's are from $-\infty$ to $+\infty$. Since the energy term in the exponent involves only the p's, integration is readily performed with respect to the q's which may take values quite independently of the p's. Integration with respect to the q's yields

$$\iiint dq_x\, dq_y\, dq_z = abc = V \qquad (7.1.3)$$

where V is the volume of the box. Equation (7.1.2) thus becomes

$$q = h^{-3}V \int\!\!\int\!\!\int_{-\infty}^{\infty} \exp\left[-\frac{p_x^2 + p_y^2 + p_z^2}{2mkT}\right] dp_x\, dp_y\, dp_z \qquad (7.1.4)$$

Since the p's may take any values independently, integration with respect to each p may be carried out without reference to the other p's. The integral (7.1.4) may then be split into three factors which are identical except for the specification of the variable. We thus obtain

$$q = h^{-3}V \left\{\int_{-\infty}^{\infty} \exp\left[-\frac{p^2}{2mkT}\right] dp\right\}^3 \qquad (7.1.5)$$

The integral is that of the Gaussian error curve with $2\sigma^2 = 2mkT$. Since the integral of this curve is $I = (2\pi\sigma^2)^{1/2}$ we obtain finally for the partition function

$$q = (2\pi mkT/h^2)^{3/2} V \qquad (7.1.6)$$

This result is identical to that obtained by the quantum formulation. This might have been expected since we showed that the description of the motion of a system in terms of a cellular phase space was in strict correspondence with the quantum description, if the boundaries of the cells were correctly assigned, and that for translational motion the energies of the quantum states were so close together that the exact positions of the boundaries become irrelevant. This is equivalent to the assumption made in Section 6.1 that the exact summation can be replaced by an integration.

7.2 THE CLASSICAL ROTATOR

One-dimensional rotator

For rotation in one dimension (or more correctly in a plane about a single axis of rotation) of a mass m constrained to move at a distance r

from a fixed axis, the energy is entirely kinetic and is given by

$$\epsilon = \frac{I\dot{\theta}^2}{2} = \frac{p_\theta{}^2}{2I} \tag{7.2.1}$$

where $\dot{\theta}$ is the angular velocity in radians s^{-1}. p_θ is the angular momentum defined as $I\dot{\theta}$. It is readily shown that equation (7.2.1) is equivalent to the equation $\epsilon = mv^2/2$ where v is the orbital speed (say in m s^{-1}). The angular momentum may take any value from $-\infty$ to $+\infty$ and the conjugate positional coordinate q_θ can have any value between 0 and 2π, corresponding to one complete revolution. The partition function is then

$$q_{\text{rot(1-D)}} = h^{-1} \int\int \exp\left[-\frac{p_\theta{}^2}{2IkT}\right] \mathrm{d}p_\theta\,\mathrm{d}q_\theta$$

$$= h^{-1}(2\pi)\int_{-\infty}^{\infty} \exp\left[-\frac{p_\theta{}^2}{2IkT}\right]\mathrm{d}p_\theta$$

$$= h^{-1}(2\pi)\,(2\pi IkT)^{1/2} = \sqrt{\pi}\,(8\pi^2 IkT/h^2)^{1/2} \tag{7.2.2}$$

Two-dimensional rotator

The dynamics of the two-dimensional rotator are considerably more complex than those of the simple rotator. The two-dimensional rotator may be represented as a mass m constrained to move on the surface of a sphere of radius r with a velocity v. The moment of inertia is $I = mr^2$. The position of the mass is determined by the two angles θ and ϕ as shown in Figure 7.1. The total kinetic energy of the moving mass is $mv^2/2$, and as shown in the figure this may be resolved into parts corresponding to velocities in two perpendicular directions. One of these component velocities involves change only in the angle θ, and the other only in ϕ. By application of Pythagoras' theorem we obtain

$$v^2 = (r\dot{\theta})^2 + (r\dot{\phi}\sin\theta)^2 \tag{7.2.3}$$

Thus the energy may be given as

$$\epsilon = (mr^2\dot{\theta}^2 + mr^2\,\dot{\phi}^2\sin^2\theta)/2 = \tfrac{1}{2}I(\dot{\theta}^2 + \dot{\phi}^2\sin^2\theta) \tag{7.2.4}$$

Now the angular momentum for rotation by change of θ in the plane containing the z-axis is simply $p_\theta = mr^2\dot{\theta} = I\dot{\theta}$. The angular momentum for rotation by change of ϕ is however

$$p_\phi = m(r\sin\theta)^2\dot{\phi} = I\sin^2\theta\dot{\phi} \tag{7.2.5}$$

$r\sin\theta$ being the distance of the mass from the z-axis.

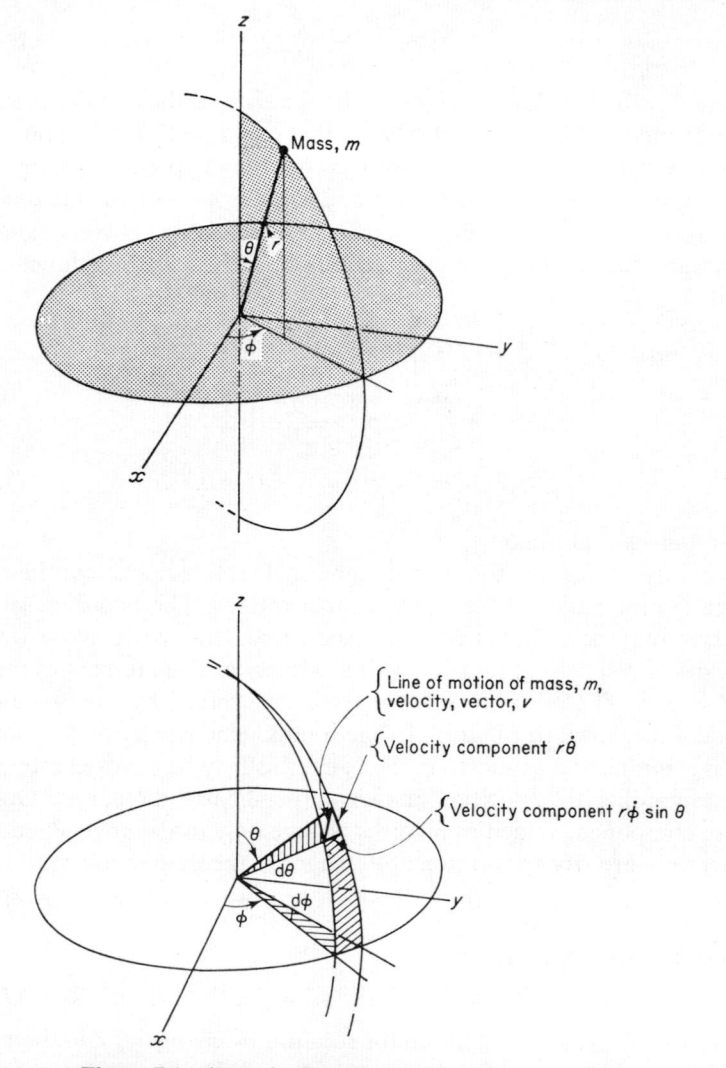

Figure 7.1 (*Upper*): Coordinates of a mass m on the surface of a sphere of radius r. (*Lower*) Velocity components of a mass m moving on the surface of a sphere of radius r.

In terms of p_θ and p_ϕ the energy is thus given as

$$\epsilon = \frac{p_\theta{}^2}{2I} + \frac{p_\phi{}^2}{2I \sin^2 \theta} = \frac{1}{2I}\left(p_\theta{}^2 + \frac{p_\phi{}^2}{\sin^2 \theta}\right) \tag{7.2.6}$$

The ranges of p_θ and p_ϕ are from $-\infty$ to $+\infty$. The ranges of the conjugate positional coordinates are $0 < \theta < \pi$ and $0 < \phi < 2\pi$. (*Note*: if the limits of θ were taken as $0 < \theta < 2\pi$, the total volume of phase space would be counted twice.) The partition function thus becomes

$$q_{\text{rot(2-D)}} = h^{-2} \int\int\int\int \exp\left[-\frac{p_\theta{}^2 + p_\phi{}^2/\sin^2\theta}{2IkT}\right] dp_\theta \, dp_\phi \, d\theta \, d\phi \tag{7.2.7}$$

Since ϕ does not appear in the exponent we can integrate immediately with respect to ϕ giving a factor 2π, that is

$$q = h^{-2}(2\pi) \int\int\int \exp\left[-\frac{p_\theta{}^2 + p_\phi{}^2/\sin^2 \theta}{2IkT}\right] dp_\theta \, dp_\phi \, d\theta \tag{7.2.8}$$

Integration may now be performed with respect to p_θ keeping p_ϕ and θ constant. The integral is that of the Gaussian error curve with $2\sigma^2 = 2IkT$, and so gives a factor $(2\pi IkT)^{1/2}$, that is

$$q = h^{-2}(2\pi)\,(2\pi IkT)^{1/2} \int\int \exp\left[-\frac{p_\phi{}^2}{\sin^2 \theta \times 2IkT}\right] dp_\phi d\theta \tag{7.2.9}$$

Integration with respect to p_ϕ keeping θ constant gives a further factor $(2\pi IkT)^{1/2} \sin \theta$. Thus

$$q = h^{-2}(2\pi)\,(2\pi IkT)^{1/2}\,(2\pi IkT)^{1/2} \int_0^\pi \sin \theta \, d\theta \tag{7.2.10}$$

The integral with respect to θ is 2, and the final result is therefore

$$q_{\text{rot(2-D)}} = 8\pi^2 IkT/h^2 \tag{7.2.11}$$

The results for both one- and two-dimensional rotators are seen to be identical to those obtained from the quantum formulation when the exact summations were replaced by integrations.

A similar but more involved dynamical argument can be applied to the three-dimensional rotator. The mathematics are fully described elsewhere and lead to the result given in equation (6.2.17) (see Rushbrooke, and Fowler & Guggenheim for example).

In the same way as allowance for the identity of systems led to the factor of $N!$ in the expression for the partition function for a gas, i.e. $Q = q^N/N!$, so the identity of equivalent configurations of a molecule which can be arrived at by rotation requires the introduction of a symmetry number in the denominator of the expressions for q_{rot}. The classical formulation thus results in equations which are identical to those arrived at by the quantum treatment.

7.3   THE CLASSICAL HARMONIC OSCILLATOR

The energy of a classical harmonic oscillator depends upon both the position and velocity (or momentum) of the oscillator. If the oscillator is regarded as a mass m constrained to move in a straight line according to simple harmonic motion, the energy is given by

$$\epsilon = fq^2/2 + p^2/2m \qquad (7.3.1)$$

where f is the force constant, and q the displacement from equilibrium. For a more complex oscillator a similar equation holds but p, q and m are now functions of the momenta, positions and masses of the individual atoms. The classical partition function is obtained by integration over all accessible regions of phase space, that is for both p and q between $-\infty$ and $+\infty$.

$$q_{vib} = h^{-1} \int \int \exp \left[-\frac{fq^2m + p^2}{2mkT} \right] dp \, dq \qquad (7.3.2)$$

Since p and q are independently variable the integrations can be performed independently giving

$$q_{vib} = h^{-1}(2\pi mkT/fm)^{1/2} (2\pi mkT)^{1/2}$$

$$= h^{-1}(2\pi kT) (m/f)^{1/2} \qquad (7.3.3)$$

Solution of the equation of motion for the oscillator

$$\dot{p} + fq = 0 \qquad (7.3.4)$$

shows that the frequency of oscillation in cycles per unit time is

$$\nu = (1/2\pi) (f/m)^{1/2} \qquad (7.3.5)$$

Substituting (7.3.5) into (7.3.3) then gives

$$q_{vib} = kT/h\nu \qquad (7.3.6)$$

This is, of course, the same result as was obtained in (6.3.14) by assuming that θ_v/T was small in the formula obtained by the quantum treatment, or by replacing the summation required in this treatment by an integration.

In general the classical partition function can always be obtained from the quantum formulation by replacing the summation demanded by the latter by an integration, or, what is equivalent, by assuming that the energy levels of the quantized system are very close together compared to kT.

PROBLEMS

7.1 In equation (7.3.4) $\dot{p}$ is the rate of change of momentum, that is $m \, d^2q/dt^2$. Examine whether solutions of the type $q = A \sin \omega t$ will satisfy the equation, and hence prove equation (7.3.5).

7.2 Why is the assumption that summation can be replaced by integration valid only when the energy levels are close together?

CHAPTER 8

THERMODYNAMIC FUNCTION FOR THE IDEAL ASSEMBLY OF LOCALIZED SYSTEMS

In an ideal assembly of localized systems the systems are imagined to be so weakly coupled that they possess well-defined private quantum states whose energies are unaffected by those of the neighbouring systems. The assembly and molecular partition functions are then, and only then, related by the equation

$$Q = q^N \tag{8.0.1}$$

N being the number of systems in the assembly. Unfortunately no real assemblies of this type exist, but atomic crystals (rare gases, metals, etc.), and a few covalent solids (diamond, silicon carbide, etc.) approximate roughly to this description especially at high temperatures, for although the systems of any crystalline solid must interact strongly, the main effect of this interaction is to maintain each system in a deep potential well from which it cannot escape. The motion of any system within its own potential well is then almost unaffected by the motions of neighbouring systems. To a first approximation each system may thus be considered as an independent oscillator.

Until about 1900 it was believed that the ratio of the molar heat capacity to the gas constant, C_v/R, was approximately three for all monatomic solids (the law of Dulong and Petit). Diamond was a striking exception to the law; its heat capacity lay well below $3R$ even at 1000 K, although it rose towards this value as the temperature increased. The anomaly was not resolved until Einstein applied quantum statistical mechanics to the problem. The success of his treatment provided one of the early justifications for the then new quantum theory.

144

8.1 THE EINSTEIN MODEL FOR A MONATOMIC CRYSTAL

Einstein proposed that a crystal of N identical atoms could be regarded as an assembly of $3N$ one-dimensional harmonic oscillators, or alternatively as an assembly of N three-dimensional isotropic oscillators. The molecular partition for the one-dimensional harmonic oscillator, when the energies of the quantum states are measured from the classical energy zero, is (see equation 6.3.7)

$$q_{\text{vib}} = \exp\left[-\theta_v/2T\right](1 - \exp\left[-\theta_v/T\right])^{-1} \qquad (8.1.1)$$

where $\theta_v = h\nu/k$, ν being the fundamental vibration frequency of the oscillator. The energy of $3N$ such systems is then

$$E = 3NkT^2 \left(\frac{\partial \ln q_{\text{vib}}}{\partial T}\right)_v$$

$$= 3NkT^2 \frac{\partial}{\partial T} \ln\left\{\exp\left[-\frac{\theta_v}{2T}\right]\left(1 - \exp\left[-\frac{\theta_v}{T}\right]\right)^{-1}\right\}$$

$$= 3NkT^2 \frac{\partial}{\partial T}\left\{-\frac{\theta_v}{2T} - \ln\left(1 - \exp\left[-\frac{\theta_v}{T}\right]\right)\right\} \qquad (8.1.2)$$

The differentiation is carried out at constant volume; this implies constant environment for each system, and therefore that ν and θ_v are constant. We obtain:

$$\bar{E} = 3NkT^2 \left\{\frac{\theta_v}{2T^2} + \frac{(\theta_v/T^2)\exp\left[-\theta_v/T\right]}{(1 - \exp\left[-\theta_v/T\right])}\right\}$$

$$= \tfrac{3}{2}Nh\nu + 3Nh\nu \exp\left[-\theta_v/T\right](1 - \exp\left[-\theta_v/T\right])^{-1} \qquad (8.1.3)$$

The first term on the right-hand side is the zero point energy of the crystal, that is the lowest energy that the crystal or assembly can have relative to the classical energy zero (see Figure 6.13). The second term is the temperature dependent part of the energy, and may be expressed as

$$\bar{E} - E_0 = 3Nh\nu\,(\exp\left[\theta_v/T\right] - 1)^{-1} \qquad (8.1.4)$$

Equations (8.1.3) and (8.1.4) may be expressed more concisely in the forms

$$\bar{E}/3NkT = x\{\tfrac{1}{2} + e^{-x}(1 - e^{-x})^{-1}\} = x\{\tfrac{1}{2} + (e^x - 1)^{-1}\}$$

$$\qquad (8.1.5)$$

$$(\bar{E} - E_0)/3NkT = x(e^x - 1)^{-1} \qquad (8.1.6)$$

where $x = \theta_v/T = h\nu/kT$.

The heat capacity at constant volume is obtained by differentiation of (8.1.4) at constant volume or constant θ_v:

$$C_v = \left(\frac{\partial \bar{E}}{\partial T}\right)_V = \left\{\frac{\partial(\bar{E} - E_0)}{\partial T}\right\}_V$$

$$= 3Nh\nu \frac{\theta_v}{T^2} \exp\left[\frac{\theta_v}{T}\right] \left(\exp\left[\frac{\theta_v}{T}\right]_* - 1\right)^{-2}$$

$$= 3Nk\left(\frac{\theta_v}{T}\right)^2 \left\{\frac{\exp[\theta_v/T]}{(\exp[\theta_v/T] - 1)^2}\right\} \tag{8.1.7}$$

Dividing both sides by $3Nk$ and replacing θ_v/T by x gives

$$C_v/3Nk = x^2 e^{-x}(1 - e^{-x})^{-2} = x^2 e^x (e^x - 1)^{-2} \tag{8.1.8}$$

Both $E/3NkT$ and $C_v/3Nk$ are dimensionless quantities which depend only upon the dimensionless parameter $x = h\nu/kT$. They are thus readily evaluated once the fundamental vibration frequency is known. Values of the functions on the right-hand sides of equations (8.1.6) and (8.1.8) are given in the tables of harmonic oscillator parameters (see, for example, Taylor and Glasstone's *Physical Chemistry*).

Figure 8.1 shows plots of $C_v/3Nk$ and $\bar{E}/3NkT$ against $1/x = T/\theta_v$. T/θ_v is called the "reduced vibrational temperature."

When T/θ_v is small, that is x large, the heat capacity can be approximated to

$$\mathop{Lt}_{T \to 0} \{C_v/3Nk\} = x^2 e^{-x} \tag{8.1.9}$$

since e^{-x} in the denominator of (8.1.8) becomes negligible compared to unity. The function on the right-hand side of (8.1.9) is similar to that in equation (4.2.55), and so tends to zero as x tends to infinity or T to zero. The limiting value of C_v at 0 K is therefore zero.

When T/θ_v is large, that is x small, the limiting value of C_v is obtained from (8.1.8). Writing e^x as $1 + x + x^2/2 + \ldots$ we obtain

$$C_v/3Nk = \frac{x^2(1 + x + x^2/2 + \cdots)}{(x + x^2/2 + x^3/3! + \cdots)^2} = \frac{(1 + x + x^2/2 + \cdots)}{(1 + x/2 + x^2/6 + \cdots)^2}$$

$$= \frac{(1 + x + x^2/2 + \cdots)}{(1 + x + (7/12)x^2 + \ldots)} = 1 - (x^2/12) + \cdots \approx 1 \tag{8.1.10}$$

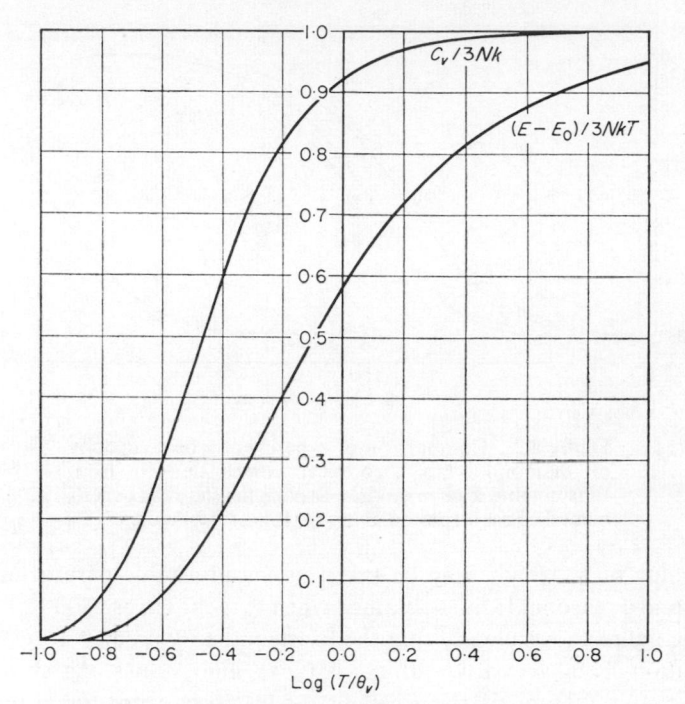

Figure 8.1 Dependence of $C_v/3Nk$, and of $(\bar{E} - E_0)/$ $3NkT$ upon T/θ_v for an assembly of localized three-dimensional harmonic oscillators.

The term $(x^2/12)$ is negligible when x is small. When the crystal contains one mole $N = N_A$, and the molar heat capacity is

$$C_v = 3N_A k \tag{8.1.11}$$

According to the law of Dulong and Petit, C_v should be close to $3R$ for monatomic solids. This agrees with the predictions of theory in the high temperature approximation if we make the identification (8.1.12).

$$k = R/N_A \tag{8.1.12}$$

Figure 8.2 compares the observed molar heat capacity of diamond with that predicted by the Einstein formula. A logarithmic temperature scale is employed so that the two curves may be fitted simply by moving one of them (say the experimental curve) along the x-axis, leaving its general shape unchanged. This manœuvre corresponds to changing

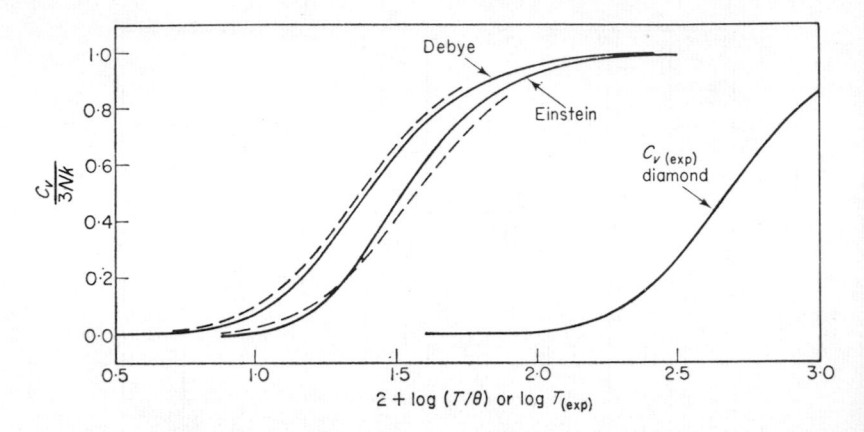

Figure 8.2 Comparison of experimental heat capacity of diamond with theoretical curves derived from Einstein and Debye models. Broken lines are the experimental curve displaced along the $\log_{10}(T/\theta_v)$-axis.

the value of θ_v which may be taken as an adjustable parameter. For diamond a reasonable fit is obtained when θ_v is chosen as 1400 K.

The figure shows clearly that the Einstein theory predicts too sharp a transition from very low values of C_v to high values. As shown by Debye, this feature of the Einstein model arises from the neglect of more complex vibrations of the crystal. These may be thought of either as coupled vibrations of the systems of the crystal now no longer considered independent, or as the normal vibrational modes of the crystal with some maximum frequency being set. The improved account of the low temperature heat capacity of monatomic solids given by the Debye theory is shown in Figure 8.2. The Debye formula for the heat capacity cannot be written in closed form, but like the Einstein formula the dependence on temperature can be expressed as a function of a reduced temperature T/θ_D, where θ_D has an analogous meaning to θ_v, and may be called the "characteristic Debye temperature" of the crystal. For diamond $\theta_D = 1860$ K for the best fit to the experimental data. At low temperatures the Debye model leads to the well known cube law approximation

$$\frac{C_v}{3Nk} = \frac{4\pi^4}{5}\left(\frac{T}{\theta_D}\right)^3, \quad T \ll \theta_D \tag{8.1.13}$$

This formula is to be compared with (8.1.10) given by the Einstein theory, which gives a value of C_v which falls away much more rapidly with temperature than does the Debye formula.

Comparative data for a large number of monatomic solids which includes metals, covalent solids and salts of monatomic ions (e.g. NaCl) is given by Fowler and Guggenheim. There is excellent agreement between the experimental data and the values of C_v predicted by Debye. For details of the Debye theory the reader is referred to Andrews' *Equilibrium Statistical Mechanics*, Wiley, 1963, and other more advanced texts.

PROBLEMS

8.1 Plot equation (8.1.9) and (8.1.13) up to $T/\theta = 0.05$ and note the difference.

8.2 Using equations (5.1.6) and (5.1.7) derive expressions for the free energy F, and the entropy S of an Einstein solid. Using $\theta_v = 1400$ K for diamond, calculate S_{Einstein} for diamond at 300 K and 1000 K. The experimental values are 2.416 J K^{-1} mol^{-1} and 19.66 J K^{-1} mol^{-1}. How do you account for the discrepancies?

8.3 The experimental molar heat capacity of KCl has been accurately determined by Southart and Nelson (*J. Amer. Chem. Soc.*, **55**, 4865 (1933)). A selection of the data are given below. Explain why the high temperature value approaches 50 J K^{-1} mol^{-1}, rather than 25 J K^{-1} mol^{-1} as expected on the basis of the Einstein or Debye theories. Plot C_p against $\log T$ and compare the curve obtained with that calculated by the Einstein and Debye formulae (see Figure 8.2). Hence evaluate θ_{Einstein} for KCl.

T/K	16.69	21.21	25.06	32.41	39.86	49.27
C_p/J K^{-1} mol^{-1}	1.79	3.52	5.48	9.87	14.78	20.61

T/K	59.61	90.80	116.47	141.17	201.32	262.94
C_p/J K^{-1} mol^{-1}	26.47	37.25	41.97	45.02	48.49	50.25

CHAPTER 9

THERMODYNAMIC FUNCTIONS FOR THE IDEAL ASSEMBLY OF NON-LOCALIZED SYSTEMS

An ideal assembly of non-localized systems when described in classical terms consists of randomly moving systems occupying the same general region of space, and having interactions so weak that the systems may be considered to behave independently. Physically this is possible only if the systems are point masses, and exert no attractive or repulsive forces on each other. When considered in quantum mechanical terms, independence implies that the same quantum states with precisely the same energies are accessible to all systems, and that the ensemble of quantum states is the same whether the assembly contains one or 10^{23} systems. No assumption is made about the motion of the systems or about their size. Indeed both of these concepts are foreign to quantum mechanics. The systems are not thought of as moving and if "size" has any meaning it is the total volume of the assembly. In spite of the apparently wide discrepancy between the classical and quantal pictures the partition function is the same whichever model we use.

For an assembly of this type the assembly and molecular partition functions are related by the equation

$$Q = q^N/N!$$

$$\ln Q = N \ln q - \ln N! = N \ln q - N \ln N + N \qquad (9.0.1)$$

We have shown in Section 5.3 that the molecular partition function may be factorized.

$$q = q_{\text{trans}} q_{\text{rot}} q_{\text{vib}} \qquad (9.0.2)$$

and in Section 6.1 that q_{trans} is proportional to the volume of the assembly. Thus

$$q = q° \times V = q°_{\text{trans}} q_{\text{rot}} q_{\text{vib}} \times V \qquad (9.0.3)$$

The assembly partition function is then

$$\ln Q = N \ln q^\circ - N \ln N/V + N \tag{9.0.4}$$

Since the ratio N/V is the concentration of systems, c,

$$\ln Q = N \ln q^\circ - N \ln c + N \tag{9.0.5}$$

The contributions from the various degrees of freedom to the thermodynamic parameters E, F and S were obtained in Section 5.3. When the volume dependence of q_{trans} is included we obtain

$$
\begin{aligned}
E_{trans} &= NkT^2(\partial \ln q_{trans}/\partial T)_V \\
&= NkT^2 (d \ln q^\circ_{trans}/dT)
\end{aligned} \tag{9.0.6}
$$

$$E_{rot, vib} = NkT^2(d \ln q_{rot, vib}/dT) \tag{9.0.7}$$

$$F_{trans} = -NkT \ln q^\circ_{trans} + NkT \ln N/V - NkT \tag{9.0.8}$$

$$F_{rot, vib} = -NkT \ln q_{rot, vib} \tag{9.0.9}$$

$$S_{trans} = (E_{trans}/T) + Nk \ln q^\circ_{trans} - Nk \ln N/V + Nk \tag{9.0.10}$$

$$S_{rot, vib} = (E_{rot, vib}/T) + Nk \ln q_{rot, vib} \tag{9.0.11}$$

By virtue of (9.0.6) E_{trans} and E_{total} are independent of the volume of the assembly. This is equivalent to the analogous result for U in classical thermodynamics. Since q° is a function only of molecular parameters and temperature (equation 6.1.7), the free energy F may be written in the form

$$
\begin{aligned}
F &= F^* + NkT \ln N/V \\
&= F^* + NkT \ln c
\end{aligned} \tag{9.0.12}
$$

where F^* is the free energy at unit concentration (one system per unit volume). Equation (9.0.12) is parallel to an equivalent equation in classical thermodynamics.

The pressure of an ideal gas was shown in equation (5.3.16) to arise only from the translational motion of the systems of the assembly. According to classical thermodynamics it is given by

$$p = -(\partial F/\partial V)_T \tag{9.0.13}$$

Insertion of equation (9.0.8) for F_{trans} immediately gives

$$p = NkT/V \quad \text{or} \quad pV = NkT \tag{9.0.14}$$

This is the molecular thermodynamic equivalent of the thermo-dynamic ideal gas equation

$$pV = nRT \tag{9.0.15}$$

where n is the number of moles, and R the ideal gas constant

$$R = 8\cdot3143 \pm 0\cdot0012 \text{ J mol}^{-1}\text{ K}^{-1} \tag{9.0.16}$$

We thus obtain the identification

$$nR = Nk \quad \text{or} \quad k = R(n/N) = R/N_A \tag{9.0.17}$$

Avogadro's number N_A is obtained by independent experiments as

$$N_A = (6\cdot02252 \pm 0\cdot00028)\ 10^{23}\text{ mol}^{-1} \tag{9.0.18}$$

whence the value of k, known as the Boltzmann constant, is

$$k = \frac{(8\cdot3143 \pm 0\cdot0012)}{(6\cdot02252 \pm 0\cdot00028)\ 10^{23}}$$

$$= (1\cdot38054 \pm 0\cdot00018)\ 10^{-23}\text{ J K}^{-1} \tag{9.0.19}$$

The major source of error in the value of k arises, surprisingly, from the uncertainty in the value of the gas constant, R, not from that in Avogadro's number. It is only at this point that it is possible to obtain the size of k, introduced in Section 4.2 to make the size of the temperature unit in molecular and classical thermodynamics the same.

The entropy of an ideal gas depends upon the concentration in a similar way to the free energy F:

$$S = S^* - Nk \ln N/V$$

$$= S^* - Nk \ln c \tag{9.0.20}$$

The Gibbs free energy is obtained as

$$G = F + pV$$

$$= F + NkT$$

$$= -NkT \ln q^\circ + NkT \ln N/V$$

$$= -NkT \ln q^\circ + NkT \ln c \tag{9.0.21}$$

The translational and rotational–vibrational contributions to G are then

$$G_{\text{trans}} = -NkT \ln q^\circ_{\text{trans}} + NkT \ln c$$

$$G_{\text{rot, vib}} = -NkT \ln q_{\text{rot, vib}} \tag{9.0.22}$$

The total free energy of a mixture of species which form an ideal gas is the sum of the free energies which the components would have if they occupied the same volume on their own. Thus

$$F_{\text{mixture}} = \sum \{-N_i kT \ln q_i^\circ + N_i kT \ln N_i/V - N_i kT\} \quad (9.0.23)$$

By analogy with classical thermodynamics (Section 3.2 at equation 3.2.26), the chemical potential for a component, i, is obtained by differentiating F with respect to N_i keeping V and T and all other N_i constant:

$$\mu_i = -kT \ln q_i^\circ + N_i kT/N_i + kT \ln N_i/V - kT$$

$$= -kT \ln q_i^\circ + kT \ln c_i \quad (9.0.24)$$

It must be noted that the chemical potential defined by (9.0.24) is the partial molar Gibbs free energy per system *not* per mole. The thermodynamic chemical potential is N_A times as large as the molecular thermodynamic equivalent.

For non-ideal gases equations (9.0.23) and (9.0.24) no longer hold since molecular partition functions are no longer clearly definable. More general methods are required for this situation which use the assembly or grand partition functions.

9.1 THE TRANSLATIONAL PROPERTIES OF THE IDEAL GAS

The molecular partition function for translational motion is

$$q_{\text{trans}} = (2\pi m kT/h^2)^{3/2} V \quad (9.1.1)$$

The internal energy is obtained by application of equation (9.0.6):

$$\bar{E}_{\text{trans}} = NkT^2 (\text{d} \ln q_{\text{trans}}^\circ/\text{d}T)$$

$$= NkT^2 \frac{\text{d}}{\text{d}T} \left\{ \frac{3}{2} \ln \frac{2\pi m k}{h^2} + \frac{3}{2} \ln T \right\}$$

$$= \tfrac{3}{2} NkT$$

The heat capacity is then

$$C_{v\,(\text{trans})} = \tfrac{3}{2} Nk \quad (9.1.3)$$

The molar heat capacity is $\tfrac{3}{2}R$ in complete agreement with experimental measurements on monatomic gases whose molecules possess only translational degrees of freedom.

The translational free energy of an ideal gas is obtained from equation (9.0.8) as

$$F_{trans} = -NkT\left\{\frac{3}{2}\ln\frac{2\pi mk}{h^2} + \frac{3}{2}\ln T - \ln\frac{N}{V} + 1\right\} \quad (9.1.4)$$

and the translational contribution to the chemical potential is

$$\mu_{trans} = -kT\left\{\frac{3}{2}\ln\frac{2\pi mk}{h^2} + \frac{3}{2}\ln T - \ln\frac{N}{V}\right\} \quad (9.1.5)$$

Replacement of N/V by p/kT according to the ideal gas law gives

$$\mu_{trans} = -kT\left\{\ln\frac{(2\pi m)^{3/2}k^{5/2}}{h^3} + \frac{5}{2}\ln T - \ln p\right\} \quad (9.1.6)$$

The translational contribution to the entropy of an ideal gas is obtained from (9.0.10) after substituting for E/T:

$$S_{trans} = Nk\left\{\frac{5}{2} + \frac{3}{2}\ln\frac{2\pi mkT}{h^2} - \ln\frac{N}{V}\right\} \quad (9.1.7)$$

This equation may be separated into parts which contain fundamental constants only, and parts which contain the molecular mass, temperature and concentration:

$$S_{trans} = Nk\left\{\frac{5}{2} + \frac{3}{2}\ln\frac{2\pi k}{h^2} + \frac{3}{2}\ln m + \frac{3}{2}\ln T - \ln\frac{N}{V}\right\} \quad (9.1.8)$$

It is convenient to convert this equation so that the molecular mass may be inserted as the molecular weight, and the concentration, C, in kmol m^{-3} (identical in value to the concentration in mol litre^{-1}). To make this conversion we replace m by M/L_A, where M is the molecular weight in amu, and N/V by CL_A, where C is the concentration in kmol m^{-3}. Thus we finally obtain

$$S_{trans} = Nk\left\{\frac{5}{2} + \frac{3}{2}\ln\frac{2\pi k}{h^2} - \frac{5}{2}\ln L_A + \frac{3}{2}\ln M + \frac{3}{2}\ln T - \ln C\right\}$$

$$= Nk\left\{\frac{3}{2}\ln\frac{M}{amu} + \frac{3}{2}\ln\frac{T}{K} - \ln\frac{C}{kmol\ m^{-3}} + 1{\cdot}3354\right\} \quad (9.1.9)$$

For many purposes it is necessary to calculate the entropy at a specific pressure. For this purpose N/V in equation (9.1.8) is replaced by p/kT using the ideal gas equation. We then obtain

$$S_{\text{trans}} = Nk \left\{ \frac{5}{2} + \frac{3}{2} \ln \frac{2\pi k}{h^2} + \ln k + \frac{3}{2} \ln m + \frac{5}{2} \ln T - \ln p \right\}$$

(9.1.10)

This equation is known as the Sackur–Tetrode equation after its originators. Using the molecular weight instead of m, and the pressure in atmospheres instead of in $N\,m^{-2}$ we obtain, since $(p/N\,m^{-2}) = (P/\text{atm}) \times 1\cdot01325 \times 10^5$,

$$S_{\text{trans}} = Nk \left\{ \frac{3}{2} \ln \frac{2\pi}{h^2 L_A} + \frac{5}{2} \ln k + \frac{5}{2} - \ln (1\cdot01325 \times 10^5) \right. $$

$$\left. + \frac{3}{2} \ln \frac{M}{\text{amu}} + \frac{5}{2} \ln \frac{T}{K} - \ln \frac{P}{\text{atm}} \right\} \quad (9.1.11)$$

Evaluating the part containing only fundamental constants gives

$$S_{\text{trans}} = Nk \left\{ \frac{3}{2} \ln \frac{M}{\text{amu}} + \frac{5}{2} \ln \frac{T}{K} - \ln \frac{P}{\text{atm}} - 1\cdot1650 \right\} \quad (9.1.12)$$

The translational entropy of an ideal gas is thus immediately obtainable from a knowledge of the molecular weight, temperature and concentration or pressure. No other parameters are required. A convincing check on the validity of molecular thermodynamics, as shown by Table 9.1, is the correlation of translational entropies of monatomic gases with entropies determined calorimetrically using the third law. Entropies calculated by the methods of molecular thermodynamics are

Table 9.1 Spectroscopic and calorimetric entropies of monatomic gases at 1 atm pressure

Gas	Molecular weight M/amu	Boiling point T/K	S_{calor} $J\,K^{-1}\,mol^{-1}$	S_{spec} $J\,K^{-1}\,mol^{-1}$
Argon	39·948	87·29	129·1 ± 0·4	129·20
Krypton	83·80	119·93	144·9 ± 0·4	145·04
Xenon	131·30	165·1	157·6 ± 0·4	157·29

often called "spectroscopic entropies" while those calculated by application of the third law using calorimetric methods are called "calorimetric entropies".

9.2   THE ROTATIONAL PROPERTIES OF THE IDEAL GAS

Internal energy and heat capacity

The rotational partition functions for r degrees of freedom given in equations (6.2.26) to (6.2.28) may be written in the general form

$$\ln q_{\text{rot}} = (r/2) \ln T - \ln \sigma + C(r) \qquad (9.2.1)$$

where $C(r)$ depends upon r and contains the moment(s) of inertia and fundamental constants. For linear molecules $r = 2$, and for rigid non-linear polyatomic molecules $r = 3$, r is further increased by one for every free internal rotation.

The molar internal energy is then

$$\bar{E}_{\text{rot}} = N_A k T^2 \left(\frac{\partial \ln q_{\text{rot}}}{\partial T} \right)_V = \frac{r}{2} N_A k T = \frac{r}{2} RT \qquad (9.2.2)$$

and the molar heat capacity at constant volume is

$$C_{v(\text{rot})} = \left(\frac{\partial \bar{E}_{\text{rot}}}{\partial T} \right)_V$$

$$= \frac{r}{2} N_A k = \frac{r}{2} R \qquad (9.2.3)$$

Theory and experiment may be compared by examining the ratio of the heat capacities at constant pressure and constant volume, $C_p/C_v \equiv \gamma$. For ideal gases by thermodynamics,

$$C_p - C_v = R \qquad (9.2.4)$$

For diatomic molecules and rigid polyatomic molecules whose vibrational degrees of freedom are unexcited, the total heat capacities are

$$C_{v(\text{total})} = C_{v(\text{trans})} + C_{v(\text{rot})}$$

$$= (r + 3)RT/2 \qquad (9.2.5)$$

Thus the ratio of heat capacities should be

$$\gamma \equiv \frac{C_p}{C_v} = \frac{C_v + R}{C_v} = \frac{r + 5}{r + 3} \qquad (9.2.6)$$

Thus for monatomic species where $r = 0$, $\gamma = 1\cdot67$ according to theory. For diatomic molecules $r = 2$, and $\gamma = 1\cdot40$. For non-linear poly-atomics $r = 3$ and $\gamma = 1\cdot33$. Table 9.2 shows the excellent agreement between experiment and theory.

Table 9.2   Heat capacity ratio C_p/C_v

Substance	Temperature K	C_p/C_v Experimental	C_p/C_v[a] Theoretical
Monatomics			
He	291	1·66	1·667
Ne	292	1·64	,,
Ar	284	1·66	,,
Kr	292	1·69	,,
Xe	292	1·67	,,
Na	750–920	1·68	,,
K	660–1000	1·64	,,
Hg	550–630	1·666	,,
Diatomics			
H_2	289	1·41	1·40
N_2	293	1·40	,,
O_2	293	1·40	,,
CO	291	1·39	,,
NO	288	1·38	,,
HCl	290–373	1·39	,,
Polyatomics			
N_2O	200	1·32	1·333
NH_3	243	1·32	,,
CH_4	193	1·33	,,
C_2H_4	179	1·32	,,

[a] Assuming vibrational modes are unexcited.

Entropy

Substitution of the expressions for q_{rot} into equation (9.0.11) gives expressions for the entropy due to rotational motion.

For one-dimensional rotation (internal rotation) we obtain

$$S_{\text{rot(1-D)}} = Nk \left\{ \frac{1}{2} + \frac{1}{2} \ln \frac{8\pi^3 IkT}{h^2} - \ln \sigma_1 \right\}$$

$$= Nk \left\{ \frac{1}{2} \ln \frac{8\pi^3 ke}{h^2} + \frac{1}{2} \ln I + \frac{1}{2} \ln T - \ln \sigma_1 \right\} \quad (9.2.7)$$

The first term contains only fundamental constants and can be evaluated as a pure number. However, it is convenient to modify this equation slightly so that the moment of inertia I may be inserted in amu Å². The conversion (see Example 6.1) is

$$\frac{I}{\text{kg m}^2} = \frac{I/\text{amu Å}^2}{6 \cdot 02252 \times 10^{46}} \tag{9.2.8}$$

The entropy is therefore given as

$$S_{\text{rot(1-D)}} = Nk \left\{ \frac{1}{2} \ln \frac{8\pi^3 ke}{h^2 \times 6 \cdot 02252 \times 10^{46}} + \frac{1}{2} \ln \frac{I}{\text{amu Å}^2} + \frac{1}{2} \ln \frac{T}{K} \right.$$
$$\left. - \ln \sigma_1 \right\}$$

$$= Nk \left\{ \frac{1}{2} \ln \frac{I}{\text{amu Å}^2} + \frac{1}{2} \ln \frac{T}{K} - \ln \sigma_1 - 0 \cdot 5220 \right\} \tag{9.2.9}$$

For two-dimensional rotation we obtain

$$S_{\text{rot(2-D)}} = Nk \left\{ 1 + \ln \frac{8\pi^2 IkT}{h^2 \sigma_2} \right\} \tag{9.2.10}$$

Gathering together the constants and making the conversion of I from kg m² to amu Å² gives

$$S_{\text{rot(2-D)}} = Nk \left\{ \ln \frac{I}{\text{amu Å}^2} + \ln \frac{T}{K} - \ln \sigma_2 - 2 \cdot 1886 \right\} \tag{9.2.11}$$

For three-dimensional rotation of a rigid polyatomic molecule the corresponding expressions are

$$S_{\text{rot(3-D)}} = Nk \left\{ \frac{3}{2} + \frac{1}{2} \ln \frac{\pi (8\pi^2 kT)^3 (ABC)}{h^6} - \ln \sigma_3 \right\} \tag{9.2.12}$$

$$S_{\text{rot(3-D)}} = Nk \left\{ \frac{1}{2} \ln \frac{ABC}{\text{amu}^3 \text{ Å}^6} + \frac{3}{2} \ln \frac{T}{K} - \ln \sigma_3 - 2 \cdot 7106 \right\} \tag{9.2.13}$$

Example 9.1 Rotational and translational entropy of N_2 and HCl at 300 K and 1 atm pressure.

The relevant molecular parameters taken from Table 6.2 are:

	N_2	HCl
Molecular weight/amu	28·0	36·45
Moment of inertia/amu Å²	8·44	1·59
Symmetry number	2	1

For N_2 the molar entropy is:

$$S_{rot} = 8\cdot314(\ln 8\cdot44 + \ln 300 - \ln 2 - 2\cdot1886)$$
$$= 8\cdot314(2\cdot133 + 5\cdot704 - 0\cdot693 - 2\cdot189)$$
$$= 8\cdot314 \times 4\cdot955$$
$$= 41\cdot20 \text{ J K}^{-1}\text{ mol}^{-1}$$

$$S_{trans} = 8\cdot314(1\cdot5 \ln 28\cdot0 + 2\cdot5 \ln 300 - \ln 1\cdot000 - 1\cdot165)$$
$$= 8\cdot314 \times 18\cdot09 = 150\cdot42 \text{ J K}^{-1}\text{ mol}^{-1}$$

whence

$$S_{tot} = 191\cdot62 \text{ J K}^{-1}\text{ mol}^{-1}$$

For N_2 at 300 K the vibrational entropy is effectively zero as $\theta_v/T = 11\cdot3$. The "total" entropy calculated above can therefore be compared with the "calorimetric entropy"

$$S_{calor} = 192\cdot2 \text{ J K}^{-1}\text{ mol}^{-1}$$

For HCl the molar entropy is

$$S_{rot} = 8\cdot314 (\ln 1\cdot59 + \ln 300 - \ln 1 - 2\cdot1886)$$
$$= 8\cdot314 \times 3\cdot979$$
$$= 33\cdot08 \text{ J K}^{-1}\text{ mol}^{-1}$$

$$S_{trans} = 8\cdot314 (1\cdot5 \ln 36\cdot45 + 2\cdot5 \ln 300 - \ln 1\cdot000 - 1\cdot165)$$
$$= 8\cdot314 \times 18\cdot488$$
$$= 153\cdot71 \text{ J K}^{-1}\text{ mol}^{-1}$$

Whence

$$S_{tot} = 186\cdot79 \text{ J K}^{-1}\text{ mol}^{-1}$$

Again for HCl, $\theta_v/T = 14\cdot4$ and the vibrational contribution to the entropy is negligible, so S_{tot} may be compared with the "calorimetric entropy" at 300 K

$$S_{calor} = 186\cdot4 \text{ J K}^{-1}\text{ mol}^{-1}$$

For both N_2 and HCl there is agreement within the experimental error of about $0.5 \text{ J K}^{-1} \text{ mol}^{-1}$ in the calorimetric entropy.

Example 9.2 The rotational entropy of dimethyl acetylene at 300 K.

We may split this into two parts, the entropy for internal rotation and the entropy for overall rotation. From Example 6.7 the moments of inertia are:

Internal rotation: $I = 1.604$ amu Å^2, $\sigma_{\text{int}} = 3$

External rotation: $ABC = 6.420 \times 150.4^2$ amu^3 Å^6, $\sigma_{\text{ext}} = 6$

$$
\begin{aligned}
S_{\text{rot (int)}} &= 8.314(0.5 \ln 1.604 + 0.5 \ln 300 - \ln 3 - 0.522) \\
&= 8.314 \times 1.467 \\
&= 12.20 \text{ J K}^{-1} \text{ mol}^{-1}
\end{aligned}
$$

$$
\begin{aligned}
S_{\text{rot (ext)}} &= 8.314\{0.5 \ln (6.420 \times 150.4^2) \\
&\qquad\qquad\qquad\qquad + 1.5 \ln 300 - \ln 6 - 2.711\} \\
&= 8.314 \times 9.996 \\
&= 83.109 \text{ J K}^{-1} \text{ mol}^{-1}
\end{aligned}
$$

$$S_{\text{rot (total)}} = 95.31 \text{ J K}^{-1} \text{ mol}^{-1}$$

Hydrogen, deuterium and hydrogen deuteride

Hydrogen and its various isotopic forms are the only molecules whose moments of inertia and boiling points are sufficiently low that their rotational partition functions differ seriously from the classical value given by equation (6.2.27) (see Table 6.2), and whose rotational heat capacities are below the classical value. With H_2 and D_2, furthermore nuclear symmetry demands that there are ortho- and para-nuclear isomers. As was shown for H_2 in Section 6.2, only odd rotational quantum states are allowed for the ortho isomer, and only even rotational quantum states for the para isomer. With deuterium the opposite holds because the D nucleus contains two fundamental particles and the total nuclear wave function must accordingly be symmetrical (not antisymmetrical) in exchange of nuclei. The ortho states, which have symmetrical nuclear spin wave functions have the even rotational quantum numbers, while the para states have the odd rotational quantum numbers. The degeneracy of the nuclear spin levels for the ortho states is six, while that of the para states is three.

"Normal" H_2 and D_2 contain the high temperature equilibrium ratio of ortho to para isomers, that is $3:1$ for H_2 and $6:3$ for D_2, and since the equilibrium is attained very slowly in the absence of a catalyst, this ratio persists even at low temperatures. The thermodynamic properties of "normal" H_2 and D_2 are therefore those of $3:1$ and $6:3$ ortho:para mixtures, respectively. However, in the presence of a catalyst, for example activated charcoal, the thermodynamic properties are different and have to be obtained from the complete nuclear/rotational partition functions for "equilibrium" hydrogen which allow for free interconversion of isomers, namely:

$$q_{rot}(H_2) = 3 \sum_{j \text{ odd}} (2j + 1) \exp\left[-j(j+1)\frac{\theta_r}{T}\right]$$

$$+ \sum_{j \text{ even}} (2j + 1) \exp\left[-j(j+1)\frac{\theta_r}{T}\right] \qquad (9.2.14)$$

$$q_{rot}(D_2) = 3 \sum_{j \text{ odd}} (2j + 1) \exp\left[-j(j+1)\frac{\theta_r}{T}\right]$$

$$+ 6 \sum_{j \text{ even}} (2j + 1) \exp\left[-j(j+1)\frac{\theta_r}{T}\right] \qquad (9.2.15)$$

For HD there is no distinction between "normal" and "equilibrium" forms as no symmetry limitations exist, and the partition function has the normal form. None of the summations can, however, be evaluated in closed form, and the thermodynamic functions for H_2, D_2 and HD must be worked out numerically. Calculated and experimental heat capacities for "normal" H_2 and D_2 and for HD are shown in Figure 9.1. Several features are notable. The heat capacity of hydrogen rises more slowly than that of deuterium; this is largely due to the higher moment of inertia and hence lower θ_r of D_2. The curve for HD rises more rapidly than either of the other two although the moment of inertia of HD is between those of H_2 and D_2; this is largely because q_{rot} for HD contains all terms while q_{rot} for the ortho or para isomers contain only alternate terms. The calculated curves for HD and D_2 show slight maxima which are clearly shown by the experimental data. Finally, there is precise agreement between the experimental data (points) and calculated values (full lines) of the rotational heat capacity. This agreement is not obtained by adjusting parameters, for all those required to evaluate the theoretical curves are obtained independently from spectroscopic data.

The agreement between theory and experiment is thus a convincing proof of the correctness of the basis of molecular thermodynamics.

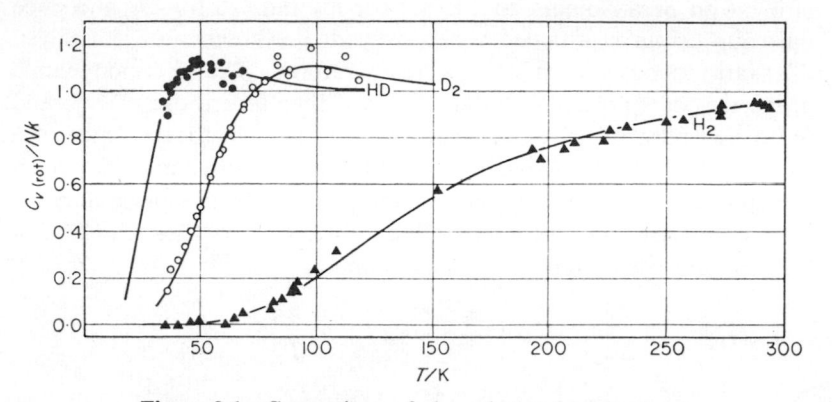

Figure 9.1 Comparison of the calculated molar heat capacities of "normal" H_2 and D_2 and of HD (full lines) with experimental data (points), after Fowler and Guggenheim.

It is also the clearest possible proof of the importance of nuclear symmetry, since the calculated curves for H_2 and D_2 would bear no relation to the experimental curves if the partition functions were obtained by summation over all rotational quantum states as is the case for HD.

The only polyatomic molecule for which a similar situation might arise is methane, which, because of its high symmetry, low moment of inertia and low boiling point, might show deviations from classical rotational behaviour. The problem is complex and is not considered here. However, Fowler and Guggenheim (Section 331) show that above about 60 K the heat capacity should exhibit the full classical value of $\frac{3}{2} Nk$. Thus the classical formulae for the rotational properties may be taken as valid above the boiling point of 95 K.

9.3 THE VIBRATIONAL PROPERTIES OF THE IDEAL GAS

The equations for the contributions to the thermodynamic functions of an ideal gas from a single vibrational mode are identical to those derived in Chapter 8 for an ideal assembly of localized systems except that $3N$ is replaced by N. This result is obtained since the vibrational quantum states are "private" to the systems of the assembly in both cases. For gases whose molecules are diatomic, and have a single vibrational mode, the energy, heat capacity, free energy and entropy are given by the formulae (9.3.1) to (9.3.4) where $\theta_v = h\nu/k$:

$$\bar{E}_{\text{vib}} = NkT\left\{\frac{\theta_v}{T}\right\}\left\{\frac{1}{2} + \frac{\exp\left[-\theta_v/T\right]}{1 - \exp\left[-\theta_v/T\right]}\right\} \tag{9.3.1}$$

$$C_{\text{vib}} = Nk\left\{\frac{\theta_v}{T}\right\}^2\left\{\frac{\exp\left[-\theta_v/T\right]}{(1 - \exp\left[-\theta_v/T\right])^2}\right\} \tag{9.3.2}$$

$$F_{\text{vib}} = NkT\left\{\frac{1}{2}\frac{\theta_v}{T} + \ln\left(1 - \exp\left[-\theta_v/T\right]\right)\right\} \tag{9.3.3}$$

$$S_{\text{vib}} = Nk\left\{\frac{(\theta_v/T)\exp\left[-\theta_v/T\right]}{1 - \exp\left[-\theta_v/T\right]} - \ln\left(1 - \exp\left[-\frac{\theta_v}{T}\right]\right)\right\} \tag{9.3.4}$$

In equations (9.3.1) and (9.3.3) a contribution arises from the zero point energy $E_{0,\text{vib}} = NkT(\theta_v/2T) = Nh\nu/2$ if the energies of the quantum states are measured from the classical zero. There is, however, no contribution from the zero-point energy to the heat capacity or to the entropy.

The equations may also be cast into dimensionless forms (9.3.5) to (9.3.8) where $x = h\nu/kT = \theta_v/T$:

$$\frac{\bar{E}_{\text{vib}} - E_{0,\text{vib}}}{NkT} = \frac{x.e^{-x}}{1 - e^{-x}} = \frac{x}{e^x - 1} \tag{9.3.5}$$

$$\frac{C_{\text{vib}}}{Nk} = \frac{x^2 e^{-x}}{(1 - e^{-x})^2} = \frac{x^2 e^x}{(e^x - 1)^2} \tag{9.3.6}$$

$$\frac{F_{\text{vib}} - E_{0,\text{vib}}}{NkT} = \ln(1 - e^{-x}) = -x + \ln(e^x - 1) \tag{9.3.7}$$

$$\frac{S_{\text{vib}}}{Nk} = x.e^{-x}(1 - e^{-x})^{-1} - \ln(1 - e^{-x})$$
$$= x.e^x(e^x - 1)^{-1} - \ln(e^x - 1) \tag{9.3.8}$$

The functions on the right-hand side of equations (9.3.5) and (9.3.6) have been plotted against $T/\theta_v = 1/x$ in Figure 8.1. Those on the right-hand sides of equations (9.3.7) and (9.3.8) are shown in Figure 9.2. The numerical values of the functions are tabulated in several places, for example Taylor and Glasstone's *Physical Chemistry*.

Figure 9.3 compares the experimental values of the molar heat capacities of a number of diatomic molecules with the theoretical curve

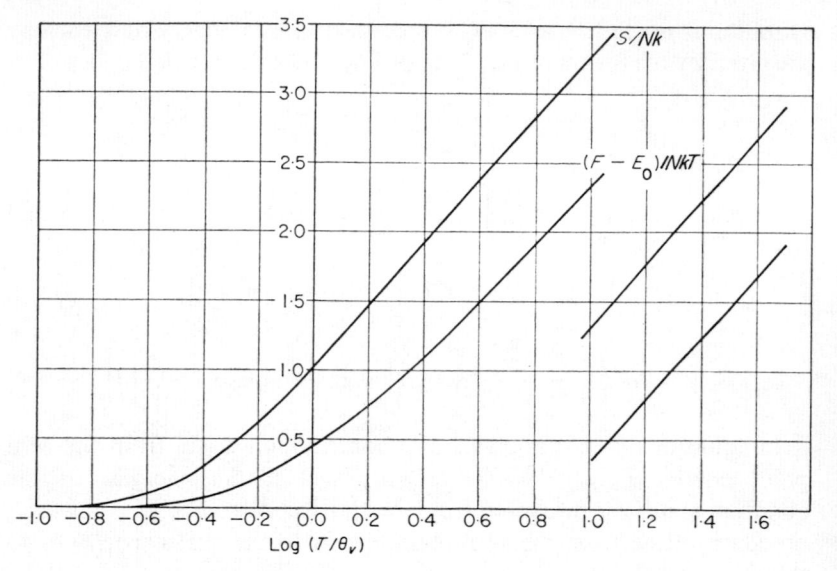

Figure 9.2 S/Nk and $(F - E_0)/NkT$ for an assembly of N harmonic oscillators as functions of (T/θ_v). Curves for $\log_{10} (T/\theta_v) > 1$ are displaced downwards by $2\cdot00$ units.

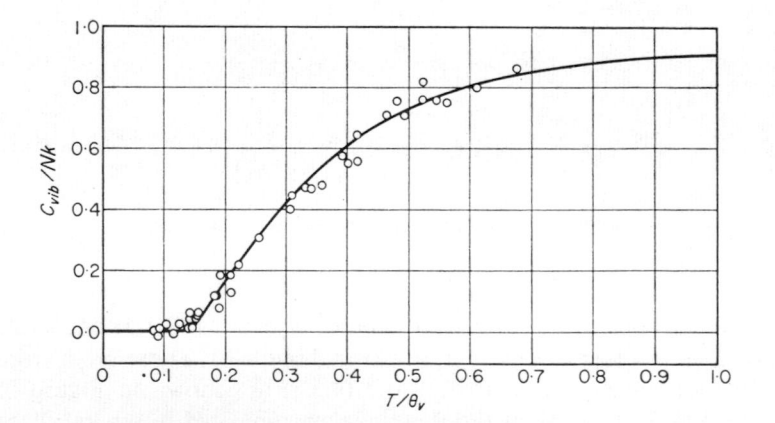

Figure 9.3 Vibrational contribution to the heat capacity of diatomic molecules compared with theory. Data included are for O_2, N_2, CO and Cl_2. Data from Fowler and Guggenheim.

of equation (9.3.6). C_{vib} is obtained from the experimental value of C_p by subtracting $\frac{7}{2}RT$.

The agreement between theory and experiment is closer here than for monatomic crystals where the assumption that the systems behave as independent harmonic oscillators is far from correct.

For polyatomic molecules the vibrational contributions to the thermodynamic functions are obtained from equations (9.3.1) to (9.3.8) by adding one term for each mode of vibration. For example, if $x_i = hv_i/kT$, the vibrational energy is

$$(\bar{E}_{vib} - E_{0,vib}) = NkT \sum \{x_i/(\exp [x_i] - 1)\} \qquad (9.3.9)$$

When the energies of the quantum states are measured from the classical energy zero and zero-point energy is

$$E_{0,vib} = \sum Nhv_i/2 \qquad (9.3.10)$$

Example 9.3 Vibrational contributions to the heat capacity and entropy of carbon dioxide at 300 and 1000 K.

Table 9.3 lists the frequencies, ω, $x = h\omega c/kT$, C_{vib}/Nk and S_{vib}/Nk

Table 9.3 Vibrational properties of carbon dioxide

Frequency ω/cm^{-1}	$x = \theta_v/T$		C_{vib}/Nk		S_{vib}/Nk	
	300 K	1000 K	300 K	1000 K	300 K	1000 K
1351 [1][a]	6·480	1·944	0·065	0·737	0·011	0·479
672·2 [2]	3·224	0·967	0·449	0·926	0·174	1·071
2396 [1]	11·491	3·447	0·001	0·404	0·000	0·146
Totals			0·964	2·993	0·359	2·767

[a] [ ] degeneracy of mode.

for each mode of vibration. The contributions from the doubly degenerate bending mode must be taken twice. From the sums given in the table we obtain:

$$C_{vib}(300 \text{ K}) = 8·314 \times 0·964$$
$$= 8·01 \text{ J K}^{-1} \text{ mol}^{-1}$$
$$C_{vib}(1000 \text{ K}) = 8·314 \times 2·993$$
$$= 24·87 \text{ J K}^{-1} \text{ mol}^{-1}$$

$$S_{\text{vib}}(300 \text{ K}) = 8\cdot314 \times 0\cdot359$$

$$= 2\cdot99 \text{ J K}^{-1} \text{ mol}^{-1}$$

$$S_{\text{vib}}(1000 \text{ K}) = 8\cdot314 \times 2\cdot767$$

$$= 23\cdot01 \text{ J K}^{-1} \text{ mol}^{-1}$$

The heat capacity over a wide range of temperature is shown in Figure 9.4 compared with the calculated curve. The agreement is within the experimental error. The values of C_{vib} calculated above are seen to fit on the curve.

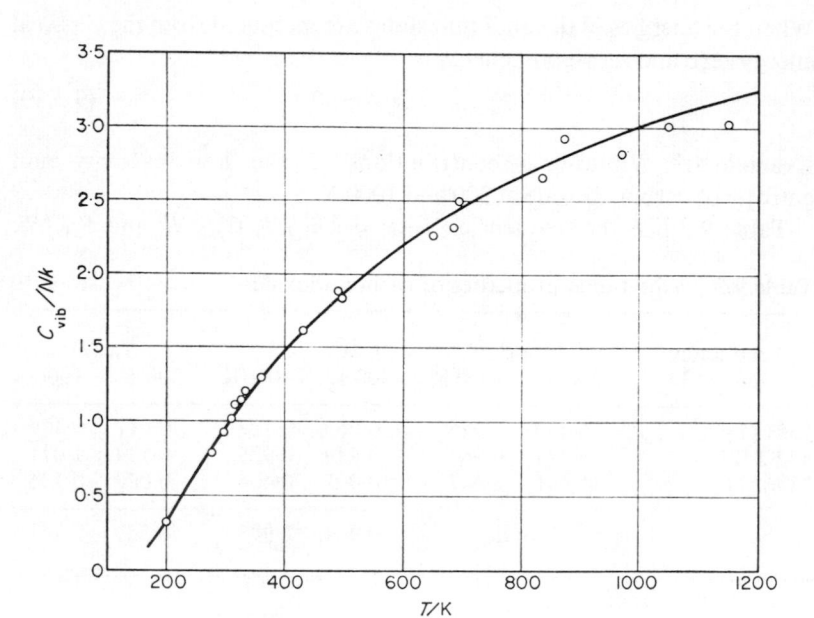

Figure 9.4 Vibrational contribution to the heat capacity of carbon dioxide compared with the theoretical curve (full line). Data from Fowler and Guggenheim.

9.4 ENTROPY CALCULATIONS FOR IDEAL GASES

With the establishment of the third law of thermodynamics around the turn of the century through the work of Nernst, Planck and others, it became possible for the first time to obtain absolute values of the entropies of chemical substances. Since the heats of formation of substances could in principle be obtained by calorimetric experiments, it now became possible to calculate free energy changes for chemical reactions, and thence equilibrium constants using the general relationships

$$\Delta G° = \Delta H° - T\Delta S° = -RT \ln K_p \qquad (9.4.1)$$

$$\Delta F° = \Delta U° - T\Delta S° = -RT \ln K_c \qquad (9.4.2)$$

Where the superscript ° implies that the change is measured between standard states (one atmosphere or unit concentration) of the reactants and products of the reaction.

Classical statistical mechanics showed how entropy could be determined but left undecided the value of an additive constant. With the advent of quantum statistical mechanics, it became possible to evaluate this constant, and to calculate entropies from molecular data. As we have seen for monatomic and diatomic molecules with high vibration frequencies the entropies obtained calorimetrically and by application of the equations of molecular thermodynamics agree within the experimental errors. The agreement between values obtained by such widely divergent experimental methods as calorimetry and spectroscopy is one of the strongest proofs of the general validity of the assumptions of molecular thermodynamics, and indeed of the whole foundation of modern theories of matter. It is therefore worthwhile to devote some space to a further comparison of entropies calculated by the two methods, and at the same time to illustrate how "spectroscopic" entropies are calculated for molecules more complex than those treated so far.

The "calorimetric" entropy of a substance is obtained by measurement of heat capacities and heats of transition. The general formula is

$$S_{\text{calor}} = S_T - S_0 = \sum \int_{T_{ij}}^{T_{jk}} \frac{C_j \, dT}{T} + \sum \frac{\Delta H_{ij}}{T_{ij}} + C \qquad (9.4.3)$$

where T_{ij} is the transition temperature from phase i to j, and ΔH_{ij} the heat of transition, and where C_j is the heat capacity of the jth phase. The first summation accounts for the entropy changes associated with heating the different phases, solid, liquid and gas, while the second summation accounts for the entropy changes associated with isothermal phase changes solid to liquid, liquid to gas, etc. The final term C is a correction term for the non-ideality of the gas, and is generally below

$1 \text{ J K}^{-1} \text{ mol}^{-1}$. It is included in order that the "calorimetric" entropy may be properly compared to the "spectroscopic" entropy which is strictly applicable only to an ideal gas. To obtain the entropy of a real gas the correction term C must be *subtracted* from the "spectroscopic" entropy.

As examples of the calculation of a "calorimetric" entropy we give the data for xenon and hydrogen chloride in Table 9.4.

Table 9.4   Data for calculation of "Calorimetric" entropies at 1 atm

T/K	Formula or method	$S^0/(\text{J K}^{-1} \text{ mol}^{-1})$
Entropy of xenon[a]		
0–10·00	Debye cube law extrapolation	3·26
10·00–161·3	Heat capacity integral: solid	59·16
161·3	Transition: solid to liquid	14·23
161·3–165·1	Heat capacity integral: liquid	3·77
165·1	Transition: liquid to vapour	76·53
	Non-ideality correction	0·58
	Total calorimetric entropy	157·5 ± 0·5
	Spectroscopic entropy	157·27
Entropy of hydrogen chloride[b]		
0–16·00	Debye cube law extrapolation	1·25
16·00–98·36	Heat capacity integral: solid I	29·54
98·36	Transition: solid I to solid II	12·09
98·36–158·91	Heat capacity integral: solid II	21·13
158·91	Transition: solid II to liquid	12·55
158·91–188·07	Heat capacity integral: liquid	9·87
188·07	Transition: liquid to vapour	85·86
	Non-ideality correction	0·42
	Total calorimetric entropy	172·7 ± 0·5
	Spectroscopic entropy	173·20

[a] From Clusius and Riccoboni, *Zeit. Phys. Chem.*, **B 38**, 81 (1938).
[b] From Giauque and Wiebe, *J. Amer. Chem. Soc.*, **50**, 101 (1928).

Table 9.5 shows the close agreement between the calorimetric and spectroscopic entropies for a number of diatomic and polyatomic molecules. The evaluation of the spectroscopic entropy of a polyatomic molecule requires knowledge of the molecular weight, the

Table 9.5 Comparison of calorimetric and spectroscopic entropies at 1 atm[a]

Molecule	T/K	$\dfrac{S^\circ_{calor}}{J\,K^{-1}\,mol^{-1}}$	$\dfrac{S^\circ_{spec}}{J\,K^{-1}\,mol^{-1}}$
N_2	298·1	192·0	191·60
O_2	298·1	205·4	205·14
HCl	298·1	186·2	186·77
HBr	298·1	199·2	198·66
HI	298·1	207·1	206·69
H_2S	212·8	194·05	194·31
NH_3	239·7	184·6	184·51
CH_4	111·5	152·8	153·18
CO_2	194·7	199·1	198·95
CS_2	318·4	240·5	241·00
SO_2	263·1	242·9	243·6
C_2H_4	169·4	198·15	198·11
CH_3Br	276·7	242·1	242·63

[a] Data from Fowler and Guggenheim. More extensive data may be found in Partington, *Treatise on Physical Chemistry*.
[b] The estimated error for most calorimetric entropies is about 0·5 J K^{-1} mol^{-1}.

moments of inertia and the fundamental vibration frequencies. Examples 9.4 and 9.5 illustrate the procedure for ethylene and methanol.

Example 9.4 Entropy of ethylene at 300 K and at 1 atm pressure.

From Example 6.2 the moment of inertia product for ethylene, the symmetry number and molecular weight are:

$$ABC = 1200\ \text{amu}^3\ \text{Å}^6, \qquad \sigma = 4, \qquad MW = 28\cdot05\ \text{amu}$$

The fundamental vibration frequencies and the characteristic vibrational temperatures for the different modes are given along with the contributions to the entropy at 300 K in Table 9.6. The contributions to the entropy are then:

$$S_{trans} = 8\cdot314\,(1\cdot5\ln 28\cdot05 + 2\cdot5\ln 300 - 1\cdot165)$$

$$= 150\cdot45\ \text{J K}^{-1}\ \text{mol}^{-1}$$

$$S_{rot} = 8\cdot314\,(0\cdot5\ln 1200 + 1\cdot5\ln 300 - \ln 4 - 2\cdot711)$$

$$= 66\cdot54\ \text{J K}^{-1}\ \text{mol}^{-1}$$

9.6 Vibrational properties of ethylene

Frequency ω/cm^{-1}	θ_v/K	$\theta_v/300$ K	S_{vib}/Nk	C_{vib}/Nk
3272	4708	15·69	0·000	0·000
3106	4469	14·90	0·000	0·000
3019	4344	14·48	0·000	0·000
2990	4302	14·34	0·000	0·000
1623	2335	7·78	0·004	0·025
1444	2078	6·92	0·008	0·047
1342	1931	6·44	0·012	0·066
1236	1778	5·93	0·019	0·094
1027	1478	4·93	0·043	0·178
943	1357	4·52	0·061	0·227
939	1351	4·50	0·062	0·230
810	1165	3·88	0·103	0·324
		Totals	0·312	1·191

Number of frequencies listed $= 12 = 3n - 6$, where $n = 6 =$ number of atoms in molecule.

$$S_{\text{vib}} = 8\cdot314 \times 0\cdot312 = 2\cdot60 \text{ J K}^{-1} \text{ mol}^{-1}$$
$$S_{\text{total}} = 219\cdot6 \text{ J K}^{-1} \text{ mol}^{-1}$$

The total spectroscopic entropy compares well with the calorimetric entropy of 220 J K^{-1} mol^{-1}.

Table 9.6 also contains the heat capacity contributions from the vibrational modes. The total molar heat capacity is then

$$C_v = 8\cdot314 \ (3\cdot00 + 1\cdot191) = 34\cdot83 \text{ J K}^{-1} \text{ mol}^{-1}$$

This compares well with the experimental calorimetric value of $35\cdot2 \pm 0\cdot5$ J K^{-1} mol^{-1}.

Example 9.5 The entropy and heat capacity of methanol vapour at 300 K and 1 atm pressure.

The moment of inertia product (Example 6.4), symmetry number and molecular weight for methanol are

$$ABC = 1740 \text{ amu}^3 \text{ Å}^6; \qquad \sigma = 1 \qquad MW = 32\cdot02 \text{ amu}$$

The vibration frequencies and parameters required for evaluation of S_{vib} and C_{vib} are listed in Table 9.7.

Table 9.7 Vibrational properties of methanol

Frequency ω/cm^{-1}	θ_v/K	$\theta_v/300$ K	S_{vib}/Nk	C_{vib}/Nk
3400	4892	16·31	0·000	0·000
2987	4298	14·32	0·000	0·000
2837	4082	13·61	0·000	0·000
1458 [4][a]	2098	6·99	0·007	0·045
1370	1971	6·57	0·011	0·061
1110 [2]	1597	5·32	0·031	0·140
1024	1473	4·91	0·044	0·180
270[b]	388	1·295	0·809	0·871
		Totals	0·954	1·572

Number of frequencies listed $= 12 = 3n - 6$, where $n = 6 =$ number of atoms in molecule.

[a] [ ] indicates degeneracy of mode. Listed contributions to S_{vib} and V_{vib} must be multiplied by degeneracy.

[b] This low frequency is for the torsional mode in methanol. The mode is more correctly described as a hindered internal rotation, the barrier is ~ 4.5 kJ mol^{-1}.

The contributions to the entropy are:

$$S_{trans} = 8.314 \, (1.5 \ln 32.04 + 2.5 \ln 300 - 1.165)$$

$$= 152.10 \text{ J K}^{-1} \text{ mol}^{-1}$$

$$S_{rot} = 8.314 \, (0.5 \ln 1740 + 1.5 \ln 300 - \ln 1 - 2.711)$$

$$= 79.61 \text{ J K}^{-1} \text{ mol}^{-1}$$

$$S_{vib} = 8.314 \times 0.954 = 7.93 \text{ J K}^{-1} \text{ mol}^{-1}$$

$$S_{total} = 239.64 \text{ J K}^{-1} \text{ mol}^{-1}$$

$$S_{calor} = 241 \pm 2 \text{ J K}^{-1} \text{ mol}^{-1}$$

The heat capacity is

$$C_v = 8.314 \, (3.00 + 1.572) = 38.01 \text{ J K}^{-1} \text{ mol}^{-1}$$

The accepted value for C_v is 35·58 J K^{-1} mol^{-1}. The discrepancy between this and our value arises from a failure to take proper account of the internal rotational mode of methanol. Had this alternatively been regarded as a free internal rotation a low value of 34·93 J K^{-1} mol^{-1} would have been obtained.

For molecules containing more than about six atoms vibrational analysis becomes increasingly difficult and unless the molecules are highly symmetrical it becomes necessary to estimate vibration frequencies. It is unfortunate that the frequencies most difficult to estimate are those associated with modes such as chain deformations in alkanes which have very low frequencies and thus contribute most to the entropy. A method of approximation which is reasonably successful for alkanes has been given by Pitzer (*J. chem. Physics*, **8**, 711 (1940)).

There are a number of important examples of marked discrepancies between the calorimetric and spectroscopic entropies. Some of these apparent anomalies are listed in Table 9.8.

Table 9.8 Calorimetric and spectroscopic entropies of substances forming imperfect crystals at 0K

Molecule	T/K	S°_{calor} $JK^{-1}mol^{-1}$	S°_{spec} $JK^{-1}mol^{-1}$	ΔS^a $JK^{-1}mol^{-1}$	g_0	$R \ln g_0{}^b$ $JK^{-1}mol^{-1}$
H_2O	298·1	185·27	188·70	3·42	$\frac{3}{2}$	3·37
D_2O	298·1	192·00	195·22	3·22	$\frac{3}{2}$	3·37
N_2O	184·6	198·15	202·92	4·77	2	5·76
CO	298·1	193·3	197·99	4·6	2	5·76
CH_3D	99·7	153·64	165·23	11·59	4	11·52
H_2	298·1	124·3	130·66	6·4	$3\frac{3}{4}$	6·85

[a] $\Delta S = S^\circ_{spec} - S^\circ_{calor}$.
[b] g_0 = degeneracy of lowest molecular energy level resulting from alternative orientations.

In all cases the spectroscopic entropy is larger than the calorimetric entropy. Equation (9.4.3) shows that the calorimetric entropy is not in fact an absolute quantity but the difference between the entropy of the substance at the temperature, T, and the entropy at absolute zero, $S_{calor} = S_T - S_0$. Thus if S_0 were finite rather than zero, S_{calor} would be less than S_T which may reasonably be identified with the spectroscopic entropy. This is broadly the accepted explanation of the anomalies. When a substance like NNO or CO is cooled and eventually solidifies the crystal is not perfect but contains almost equal numbers of NNO or CO molecules in each possible orientation, that is NNO and ONN, or CO and OC. Both molecules are sufficiently non-polar that there is no preferred orientation of one molecule relative to its neighbours. The molecular energy levels are thus doubly degenerate in the crystal, and the molecular partition function for the crystal is twice as large as might have been expected were all the molecules orientated in the same direction. The degeneracy of the lowest energy level of the crystal as a whole is thus 2^N where N is the number of systems in the assembly or crystal. The spectroscopic entropy of the crystal at 0 K is then

$$S_0 = Nk \ln 2 \quad (\text{CO and N}_2\text{O})$$

A similar situation arises for CH_3D, where there are four possible orientations of each molecule which will be equally probable when the crystallization takes place. Thus

$$S_0 = Nk \ln 4 \quad (\text{CH}_3\text{D})$$

Water is a special case where owing to the uncertainty as to the positions of the H nuclei relative to the O atoms there is a certain randomness in the structure. Pauling has shown that this should result in an entropy at 0 K of

$$S_0 = Nk \ln \tfrac{3}{2} \quad (\text{H}_2\text{O and D}_2\text{O})$$

Hydrogen provides another example where ortho- and para-molecules are randomly distributed throughout the crystal. The entropy associated with this crystal at 0 K is

$$S_0 = \tfrac{3}{4} Nk \ln 3$$

Table 9.8 shows that these explanations give a good account of the discrepancies between S_{calor} and S_{spec}.

The general conclusion is reached that wherever the calorimetric and spectroscopic data are of sufficient precision there is excellent agreement between the two methods of determining the absolute entropy of any substance. In practice calorimetric experiments are more difficult and less accurate than spectroscopic experiments, and the latter give far more detailed information about the molecules in question. It has therefore become customary to evaluate entropies and heat capacities by the molecular thermodynamic method wherever possible.

9.5   HINDERED INTERNAL ROTATION

A complication of the partition function which cannot be ignored because of its widespread practical importance arises from hindered internal rotation, which occurs with molecules such as ethane and most organic molecules. As shown in Section 6.2 the potential energy of the ethane molecule depends upon the angle of rotation of one methyl group relative to the other. It is usually assumed to obey the equation

$$V(\phi) = V_0 \sin^2(n\phi/2) = (V_0/2)(1 - \cos n\phi) \qquad (9.5.1)$$

where $n = 3$ for ethane, ϕ is the angle of one methyl group relative to the other chosen so that $\phi = 0$ when the groups are staggered, and V_0 is the height of the potential barrier to internal rotation. The form of the potential function is shown in Figure 9.5 where it is plotted over one complete cycle $(0 < \phi < 2\pi/n)$. At low energies the form of $V(\phi)$ closely approaches that of the harmonic oscillator, as can be seen by expanding (9.5.1) as a series and taking only the first term:

$$V(\phi) \approx V_0(n\phi)^2/4 \qquad \text{(low } \phi) \qquad (9.5.2)$$

If the total energy, ϵ, of the system is low its motion will approximate to harmonic with a frequency which can be shown to be

$$\nu = (n/2\pi)(V_0/2I)^{1/2} \qquad (9.5.3)$$

where I is the reduced moment of inertia for the motion.

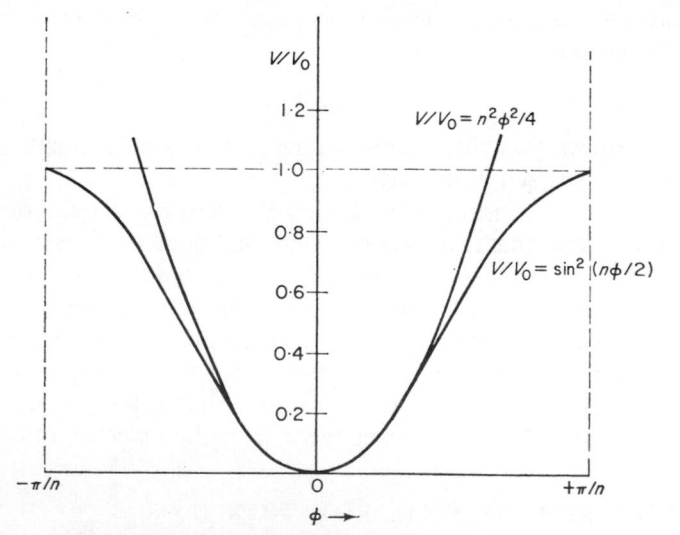

Figure 9.5 Potential function for hindered internal rotation compared with that of the harmonic oscillator with the same force constant in the limit of zero displacement.

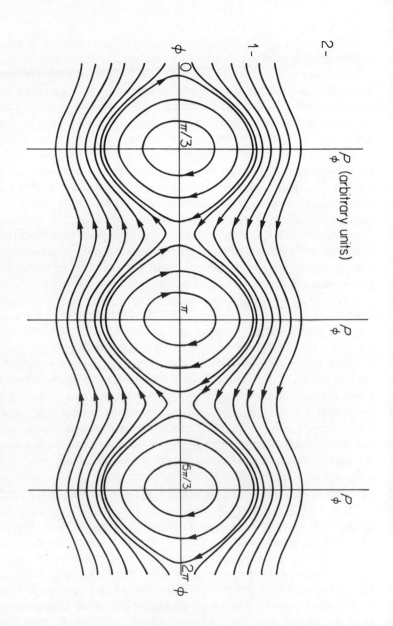

Figure 9.6 Phase space trajectories for restricted rotator. The units of p_ϕ are arbitrary.

When equation (9.5.1) is inserted into the Schrödinger equation the result after substitution and rearrangement can be written in the form of the Mathieu equation, to which solutions are well known. A conceptually simpler procedure, although ultimately an inadequate one, is to apply the more elementary ideas of the old quantum theory by constructing classical phase space trajectories appropriate to the potential function (9.5.1) for various total energies, ϵ, and then to select a suitable set so that each trajectory is separated from its neighbour by an area h. Each trajectory then corresponds to a quantum state of the system.

Figure 9.6 shows such a set of trajectories for a system having $n = 3$. The diagram shows three identical regions corresponding to the three possible equivalent internal orientations of such a molecule (see for example Figure 6.11). For low values of ϵ the motion is oscillatory and close to harmonic so the trajectories are close to elliptical. As ϵ increases the potential function deviates from the harmonic form, and the trajectories become distended. For each oscillatory energy there are three equivalent trajectories arising from the three equivalent configurations of the system which can be arrived at by internal rotation. Each such energy level has threefold degeneracy. When ϵ exceeds V_0 there is no restriction to internal rotation, and the trajectories become unbounded in the ϕ direction, although the angular momentum p_ϕ is still strongly dependent upon ϕ. However as ϵ increases still further the effect of the potential barrier diminishes and at very high total energies the rotation is essentially uniform. For each rotational energy there are two trajectories corresponding to clockwise and anticlockwise rotation: each rotational state is twofold degenerate.

According to the semi-classical picture there is a sharp transition from the oscillatory to the rotational states as ϵ passes through the value V_0. In reality the transition is much more gradual, and the full quantum treatment reveals that for ϵ close to V_0, there is considerable mixing of oscillatory and rotational states, an effect often covered by the term tunnelling. The difference between the simple phase space picture and that given by the full quantum treatment is shown in Figure 9.7 which compares the energies of the quantum states derived by the two methods. Also shown are the vibrational and rotational energies which would be observed if the system behaved either as a pure harmonic oscillator obeying (9.5.3) or as a pure rotator with a zero energy barrier.

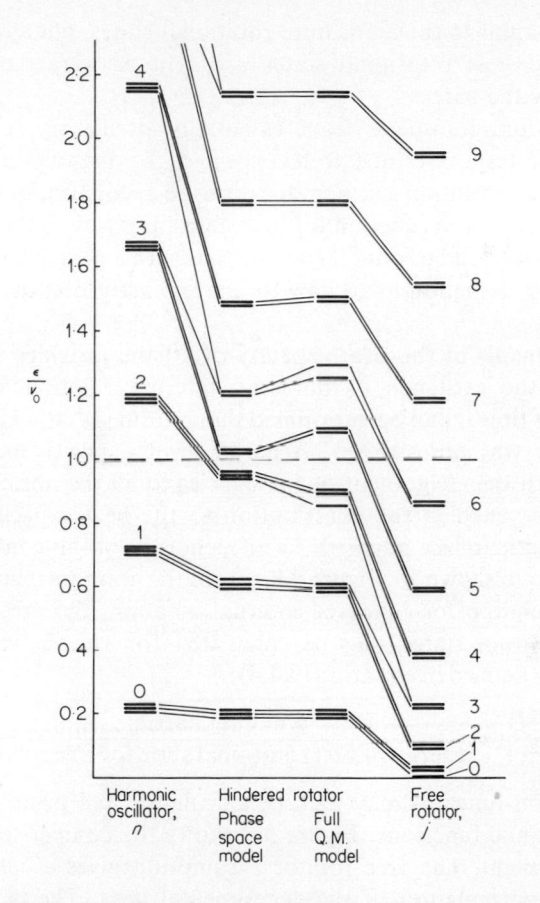

Figure 9.7 Energies of states of restricted rotator, evaluated by semi-classical and full quantum treatments, compared with those of the harmonic oscillator and free internal rotator.

In the full quantum treatment there is in fact no clear distinction between the oscillatory and rotational states. At low energies the states fall into groups of three as expected for the oscillatory states, and their wave functions are close to those for a harmonic oscillator, but as ϵ becomes close to V_0 the vibrational states are split and the uneven spacing given by the semi-classical treatment around V_0 is smoothed out. For energies well above V_0 the states appear in pairs with wave functions similar to those for pure rotational states, but as ϵ is reduced towards V_0 these rotational states are split as were the oscillatory states below the barrier.

This gradual transition from essentially oscillatory to essentially rotational states is of course to be expected, for quantum mechanics in fact makes no mention of, nor does it even recognize, different types of motion. The wave function for any state simply gives the probability that the system will be found at any instant with a particular value of ϕ: it gives no information as to how the system arrived at that configuration.

Measurements of the heat capacity of ethane provided the first evidence for the existence of hindered internal rotation about 1930. Prior to this time it had been assumed that internal rotation in molecules like ethane was unrestricted. Accurate heat capacity measurements coupled with the assignment of frequencies to all the normal modes in the molecule enabled the contribution to the heat capacity from the torsional mode to be evaluated. The dependence of this contribution on temperature is shown in Figure 9.8. From the accurate quantum mechanical calculation for hindered internal rotation, the precise energies of the quantum states may be calculated for various values of the parameter θ defined in equation (9.4.4):

$$\theta = V_0 \left\{ \frac{8\pi^2 I}{h^2} \right\} = \frac{\text{potential barrier}}{\text{energy of first rotational state for free rotator}} \quad (9.5.4)$$

The partition function may then be calculated, and hence the various thermodynamic functions. Figure 9.8 shows the comparison of theory with experiment. The free rotator assumption gives a completely inadequate interpretation of the experimental data. The best fit which can be obtained using the harmonic oscillator approximation is with $\theta_v = 350$ K or $\omega = 243$ cm^{-1}, but the experimental data deviate in a systematic way from the calculated curve. The fit assuming hindered internal rotation is excellent. The curve was calculated with $V_0/k = 1570$ K, or $V_0 = 13 \cdot 1$ kJ mol^{-1}. This value of V_0 implies an oscillation frequency for low energies of 323 cm^{-1}, slightly higher than that required to give the best fit with the harmonic oscillator approximation alone.

Tables for calculating contributions to the thermodynamic properties of gases from hindered internal rotational modes are given in several

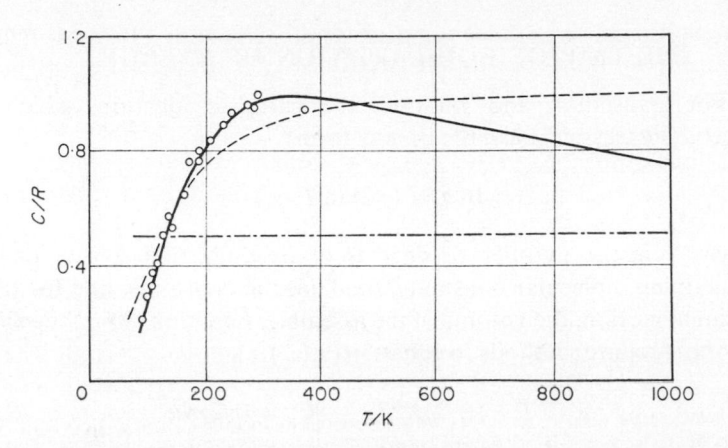

Figure 9.8 Contributions to the heat capacity of ethane from the torsional mode. Points are experimental data after Fowler and Guggenheim. ―·―·―·― C_v/R assuming free internal rotation, ――――― C_v/R for harmonic oscillator with $\omega = 243$ cm^{-1}, ――――― C_v/R for hindered internal rotation with $V_0 = 13 \cdot 1$ kJ mol^{-1}.

places (e.g. in Benson, *Thermochemical Kinetics*, Wiley, New York, 1968, and Herzberg *Infra-red and Raman Spectra of Polyatomic Molecules*, van Nostrand, New York, 1945). Typical values of barriers to internal rotation in organic molecules are given in Table 9.9.

Table 9.9   Barriers to internal rotation in organic molecules[a]

Molecule and bond	Barrier V_0 $\overline{\text{kJ mol}^{-1}}$	Molecule and bond	Barrier V_0 $\overline{\text{kJ mol}^{-1}}$
CH_3-CH_3	12·0	CH_3-OH	4·5
$CH_3-C_2H_5$	11·7	CH_3-O-CH_3	11·4
CH_3-phenyl	0·0	CH_3-CHO	4·9
$CH_3-CH=CH_2$	8·4	$CH_3-CO-CH_3$	3·3
$\genfrac{}{}{0pt}{}{CH_3}{CH_3}{>}C=CH_2$	9·3	CH_3-COOH	2·0
$CH_3C\equiv C-CH_3$	0·0	CH_3-NH_2	8·1
CH_3-CF_3	14·6	$CH_3-NH-CH_3$	13·7
CF_3-CF_3	18·2	$CH_3-N(CH_3)_2$	18·5
CCl_3-CCl_3	45·2		

[a] Data from Dale, *Tetrahedron*, **22**, 3373 (1966).

9.6 THE LAW OF EQUIPARTITION OF ENERGY

For translation and rotation, the partition functions which are essentially classical have the general form:

$$\ln q = (r/2) \ln T + f(r) \tag{9.6.1}$$

where r is the number of degrees of freedom, and f is a function containing molecular constants, fundamental constants, and for translational motion, the volume of the assembly. Equation (9.6.1) then leads by the standard methods to equation (9.6.2):

$$\bar{E} = (r/2)NkT, \qquad C_v = (r/2)Nk \tag{9.6.2}$$

This relationship is in fact general for all degrees of freedom in which the motion may be regarded as "classical" if r is identified not with the number of degrees of freedom, but with the number of so-called "square terms" which are required to give the total energy of a typical system of the assembly. It is called the "law of equipartition of energy".

For instance, the energies of translation in three dimensions, rotation in two dimensions, and vibration in one dimension are given by equations (9.6.3) to (9.6.5):

$$\epsilon_{\text{trans}} = \frac{p_x{}^2 + p_y{}^2 + p_z{}^2}{2m} \tag{9.6.3}$$

$$\epsilon_{\text{rot}} = \frac{p_\theta{}^2 + p_\phi{}^2/\sin^2 \theta}{2I} \tag{9.6.4}$$

$$\epsilon_{\text{vib}} = \frac{\mu v^2}{2} + \frac{fx^2}{2} \tag{9.6.5}$$

The number of "square terms" in these expressions are respectively three, two and two. We saw in Section 7.3 that the classical vibrational partition function was $q_{\text{vib}} = kT/h\nu$. This leads to an internal energy of NkT, and a heat capacity of Nk, in accordance with equation 9.6.2 if r is taken as two.

The general law arises from the expression for the classical partition function when the energy of any system may be expressed as the sum of independent square terms, that is when ϵ can be expressed by equation (9.6.6):

$$\epsilon = a_1 p_1{}^2 + a_2 p_2{}^2 + \cdots + a_n p_n{}^2 + b_1 q_1{}^2 + b_2 q_2{}^2 + \cdots + b_m q_m{}^2$$
(9.6.6)

where the a's and b's are constants and the p's and q's generalized momenta and positional coordinates. In this expression there are not necessarily the same number of p^2 and q^2 terms, see for example equations (9.6.3) to (9.6.5).

The classical partition function has the form

$$q = h^{-n} \int \ldots \int \exp\left[-\frac{\epsilon}{kT}\right] dp_1\, dp_2\, \ldots\, dp_n\, dq_1\, dq_2 \ldots dq_m \ldots dq_n$$
(9.6.7)

Since the p's and q's may be varied independently the integrals may be evaluated independently, for the exponential factor may itself be written as a sequence of factors in which the individual p's and q's appear independently. We can integrate immediately with respect to those q's which have no influence on the energy ϵ, that is with respect to q_{m+1} to q_n. These will be the q's for translations and rotations. This integration will produce a volume which we can call $V_{(n-m)}$. The integral now becomes

$$q = h^{-n} V_{(n-m)} \int \ldots \int \exp\left[-\frac{\epsilon}{kT}\right] dp_1\, dp_2 \ldots dp_n\, dq_1\, dq_2 \ldots dq_n$$
(9.6.8)

The integration with respect to each p or q may be carried out in succession, each integral being of the general form

$$I_i = \int_{-\infty}^{\infty} \exp\left[-\frac{a_i p_i{}^2}{kT}\right] dp_i = \left(\frac{\pi kT}{a_i}\right)^{1/2}$$
(9.6.9)

The total partition function then has the form

$$q = h^{-n} V_{(n-m)} (\pi kT)^{r/2} \{(a_1\, a_2 \ldots a_n)\, (b_1 b_2 \ldots b_m)\}^{1/2}$$
(9.6.10)

where $r = n + m$ is the number of square terms in the expression for ϵ.

The classical molecular partition function thus obeys equation (9.6.1) and the internal energy is given by (9.6.2) whenever the total energy of the typical system can be expressed by (9.6.6).

In practice, the equipartition law applies accurately only to the translational and rotational degrees of freedom because quantum effects are always important where vibrational degrees of freedom are concerned.

PROBLEMS

9.1 Check the constants in equations (9.1.9) and (9.1.12).

9.2 Calculate the translational entropies of Ar, Kr, and Xe at 300 K and 1 atm pressure. Check your result with any standard compilation of thermodynamic data.

9.3 Derive expressions for the translational contributions to the chemical potential of an ideal gas along the lines of equation (9.1.12) starting from equation (9.1.6).

Calculate the molar chemical potentials of Ar, Kr and Xe at 300 K and 1 atm pressure.

9.4 Calculate the translational, rotational and vibrational entropies of I_2 and HI in the gaseous state at 500 K and 1 atm pressure. The rotational and vibrational constants of I_2 and HI are given in Table 6.2. Compare your values with those given in standard tables.

9.5 Calculate the entropy of gaseous BF_3 at 300 K. The B—F bond length is 1·295 Å; the fundamental vibration frequencies with degeneracies in brackets are: 888, 691, 480 (2) and 1446 (2) cm^{-1}.

9.6 Calculate the molar heat capacities, C_v, of gaseous I_2 and HI at 500 K. Necessary molecular data are given in Problem 9.4.

9.7 Calculate the molar heat capacity of ammonia gas at 1000 K. The fundamental frequencies and degeneracies (in brackets) are: 3336, 950, 3414 (2), 1628 (2) cm^{-1}.

9.8 NO has a low lying electronic state 121 cm^{-1} above the ground state. The bond lengths and vibration frequencies are the same for both states. Show that the electronic partition function can then be factorized out of the complete molecular partition function, and write it down. Derive expressions for E_{el}, $C_{v(el)}$ and S_{el}, and calculate their numerical values at 300 K.

9.9 The electronic partition function for NO has the form $q_{el} = 1 + \exp[-\theta_{el}/T]$, or twice this value if electronic degeneracy is allowed for. Show that $C_{v(el)} = Nk(\theta_{el}/T)^2 \exp[\theta_e/T](1 + \exp[\theta_e/T])^{-2}$. Thence show that $C_{v(el)}$ is zero at $T = 0$ and $T = \infty$, and so must pass through a maximum at some intermediate temperature. By differentiating the equation for $\ln C_{v(el)}$ obtain an equation for the condition for the maximum, and solve it for the θ_{el}/T either numerically or by graphical means. Hence find the temperature at which the electronic molar heat capacity of NO is a maximum if $\theta_{el} = 174$ K.

9.10 Refer to the result of Problem 5.7. What happens qualitatively speaking to the mean energy of the electrons in sodium if the temperature of the metal is raised by a small amount. Does this explain the absence of any electronic heat capacity?

9.11 By reference to Benson's *Thermochemical Kinetics* or Herzberg's *Infra-red and Raman Spectra of Polyatomic Molecules*, determine the contribution to the entropy and molar heat capacity of ethane from the hindered internal rotational mode of motion at 300 K. What would be the contributions if the rotation were free, and if the barrier were extremely high? The barrier to internal rotation is $12 \cdot 0$ kJ mol^{-1}. For the reduced moment of inertia see worked Example 6.1. Compare your result with Figure 9.8.

CHAPTER 10

THE MAXWELL–BOLTZMANN DISTRIBUTION LAW

The Maxwell–Boltzmann law for distribution of molecular velocities or momenta is one of the most important results of classical statistical mechanics, and was derived in the latter part of the nineteenth century. It is directly obtained from the fundamental assumption of molecular thermodynamics which gives rise to equation (10.0.1) for the probability that a molecule of an ideal gas possesses momenta in the ranges p_x to $p_x + dp_x$, p_y to $p_y + dp_y$ and p_z to $p_z + dp_z$, and coordinates in the ranges q_x to $q_x + dq_x$, q_y to $q_y + dq_y$ and q_z to $q_z + dq_z$

$$p(p_x, p_y, p_z, q_x, q_y, q_z)\, dp_x\, dp_y\, dp_z\, dq_x\, dq_y\, dq_z$$

$$= \frac{\exp\,[-\epsilon/kT]\, dp_x\, dp_y\, dp_z\, dq_x\, dq_y\, dq_z}{q} \tag{10.0.1}$$

where q is the classical partition function, that is the integral of the numerator over all regions of accessible phase space. For the ideal gas ϵ depends only upon the p's (see equation 9.6.3) and we can accordingly integrate (10.0.1) over all positions to obtain a law for the distribution of momenta irrespective of position. Thus

$$p(p_x, p_y, p_z)\, dp_x\, dp_y\, dp_z$$

$$= \frac{\exp\,[-\epsilon/kT]\, dp_x\, dp_y\, dp_z \iiint dq_x\, dq_y\, dq_z}{\iiiiii \exp\,[-\epsilon/kT]\, dp_x\, dp_y\, dp_z\, dq_x\, dq_y\, dq_z} \tag{10.0.2}$$

Since the integral in the numerator is simply the volume of the container, and since it can furthermore be factorized out of the denominator, for again ϵ does not depend upon the q's, we obtain for the distribution of momentum irrespective of position:

$$p(p_x, p_y\, p_z)\, dp_x\, dp_y\, dp_z$$

$$= \frac{\exp\,[-(p_x^2 + p_y^2 + p_z^2)/2mkT]\, dp_x\, dp_y\, dp_z}{\iiint \exp\,[-(p_x^2 + p_y^2 + p_z^2)/2mkT]\, dp_x\, dp_y\, dp_z} \tag{10.0.3}$$

The integral in the denominator factorizes into three formally identical

184

integrals of the Gaussian form with $2\sigma^2 = 2mkT$. Each integral is thus $(2\pi mkT)^{1/2}$ giving finally

$$p(p_x, p_y, p_z)\, \mathrm{d}p_x\, \mathrm{d}p_y\, \mathrm{d}p_z$$

$$= \frac{\exp\left[-(p_x^2 + p_y^2 + p_z^2)/2mkT\right]\mathrm{d}p_x\, \mathrm{d}p_y\, \mathrm{d}p_z}{(2\pi mkT)^{3/2}} \quad (10.0.4)$$

This is the Maxwell–Boltzmann law for the distribution of momenta. We recall that the quantity $p(p_x, p_y, p_z)$ means the probability per unit range of the three momentum coordinates at values around p_x, p_y and p_z. To make the transition to the velocity distribution function we use the equation

$$p(v_x) = p(p_x) \times m \quad (10.0.5)$$

Whence

$$p(v_x, v_y, v_z)\, \mathrm{d}v_x\, \mathrm{d}v_y\, \mathrm{d}v_z = \frac{\exp\left[-(v_x^2 + v_y^2 + v_z^2)\, m/2kT\right]\mathrm{d}v_x\, \mathrm{d}v_y\, \mathrm{d}v_z}{(2\pi kT/m)^{3/2}}$$

$$(10.0.6)$$

Forms (10.0.4) and (10.0.6) lead to a number of important subsidiary results.

10.1 THE VELOCITY DISTRIBUTION IN ONE DIMENSION ONLY

If one is not interested in the velocities in the y and z directions the momentum distribution in the x-direction may be obtained by integrating (10.0.4) over all values of p_y and p_z. The integrals may be evaluated independently as was done in integrating the denominator of (10.0.3). The integration will produce two identical factors $(2\pi mkT)^{1/2}$ which will cancel out with $(2\pi mkT)$ to give finally

$$p(p_x)\, \mathrm{d}p_x = \frac{\exp\left[-p_x^2/2mkT\right]\mathrm{d}p_x}{(2\pi mkT)^{1/2}} \quad (10.1.1)$$

The appropriate velocity distribution function is

$$p(v_x)\, \mathrm{d}v_x = \frac{\exp\left[-mv_x^2/2kT\right]\mathrm{d}v_x}{(2\pi kT/m)^{1/2}} \quad (10.1.2)$$

The form of this velocity distribution function is shown in Figure 10.1. It is the well-known Gaussian error curve, and is symmetrical about

$v_x = 0$ as would be expected intuitively. The most probable molecular velocity is zero.

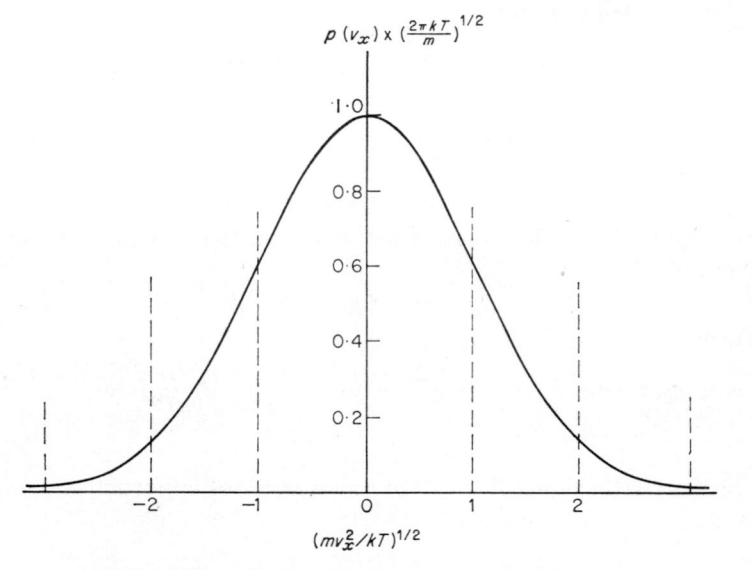

$p\,(v_x) \times \left(\frac{2\pi kT}{m}\right)^{1/2}$

$(mv_x^2/kT)^{1/2}$

Figure 10.1 Velocity distribution function for one-dimensional translation.

10.2 SPEED DISTRIBUTION IRRESPECTIVE OF DIRECTION

For many chemical purposes it is not the velocity of a molecule which is critical, but its speed. The idea of velocity contains not only speed but also direction. The speed, s, is obtained by vector addition of the velocities in the three perpendicular directions, that is:

$$s^2 = v_x^2 + v_y^2 + v_z^2 \tag{10.2.1}$$

Equation (10.2.1) represents a sphere in velocity space of radius s. For all points on the surface of a sphere of this radius the speed is constant. We can therefore integrate (10.0.6) without reference to direction by replacing the volume element $dv_x\,dv_y\,dv_z$ by a new element, namely the volume of a spherical shell of radius s and thickness ds. The appropriate replacement is therefore represented as

$$dv_x\,dv_y\,dv_z \to 4\pi s^2\,ds \tag{10.2.2}$$

The speed distribution function from (10.0.6) is then

$$p(s) \, \mathrm{d}s = (4\pi s^2) \, (m/2\pi kT)^{3/2} \exp \left[-ms^2/2kT\right] \mathrm{d}s \qquad (10.2.3)$$

The form of this function is shown in Figure 10.2. The most probable speed is now not zero. It may be found by differentiating $p(s)$ in (10.2.3)

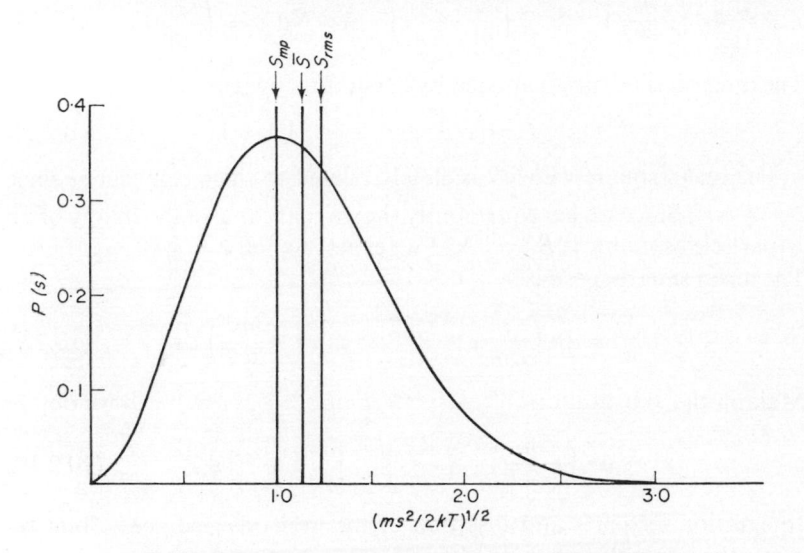

Figure 10.2 Speed distribution function for three-dimensional translation showing most probable speed, average speed and root mean square speed.

with respect to s, and setting the differential equal to zero. Ignoring constants which will cancel we then obtain:

$$0 = \left(\frac{\mathrm{d}}{\mathrm{d}s}\right) \left\{ s^2 \exp \left[-\frac{ms^2}{2kT}\right] \right\} = 2s \exp \left[-\frac{ms^2}{2kT}\right] - s^2 \frac{ms}{kT} \exp \left[-\frac{ms^2}{2kT}\right]$$

$$(10.2.4)$$

Whence cancelling out factors common to both terms we obtain

$$s_{\mathrm{mp}} = (2kT/m)^{1/2} \qquad (10.2.5)$$

The average speed which is different from the most probable speed is obtained using equation 4.2.22 as:

$$\bar{s} = \int_0^\infty s \cdot p(s) \, \mathrm{d}s = 4\pi \left(\frac{m}{2\pi kT}\right)^{3/2} \int_0^\infty s^3 \exp \left[-\frac{ms^2}{2kT}\right] \mathrm{d}s \quad (10.2.6)$$

Now

$$s^3 \, \mathrm{d}s = 2\left(\frac{kT}{m}\right)^2 \times \frac{ms^2}{2kT} \times \mathrm{d}\frac{ms^2}{2kT} = 2\left(\frac{kT}{m}\right)^2 \times X \, \mathrm{d}X$$

where $X = ms^2/2kT$. Equation (10.2.6) thus simplifies to

$$\bar{s} = 4\pi \left(\frac{m}{2\pi kT}\right)^{3/2} \times 2 \left(\frac{kT}{m}\right)^2 \times \int_0^\infty X e^{-X} \, \mathrm{d}X = \left(\frac{8kT}{\pi m}\right)^{1/2} \quad (10.2.7)$$

The integral is readily evaluated by integration by parts:

$$I = [-X.e^{-X} - e^{-X}]_0^\infty = 1 \quad (10.2.8)$$

The mean square velocity is closely related to the mean energy since $\bar{\epsilon} = \frac{1}{2} m\overline{s^2}$. Since we have previously shown that the average energy of an N-particle assembly is $\bar{E} = \frac{3}{2} NkT$ we expect to find $\bar{\epsilon} = \frac{3}{2} kT$. The mean square speed is:

$$\overline{s^2} = 4\pi \left(\frac{m}{2\pi kT}\right)^{3/2} \int_0^\infty s^4 \exp\left(-\frac{ms^2}{2kT}\right) \mathrm{d}s \quad (10.2.9)$$

Making the substitution $Y^2 = ms^2/2kT$ or $s = (2kT/m)^{1/2} Y$ we obtain

$$\overline{s^2} = 4\pi \left(\frac{m}{2\pi kT}\right)^{3/2} \left(\frac{2kT}{m}\right)^{5/2} \int_0^\infty Y^4 e^{-Y^2} \, \mathrm{d}Y \quad (10.2.10)$$

Integration by parts and insertion of the limits as one goes along reduces the integral eventually to

$$I \equiv \int_0^\infty Y^4 e^{-Y^2} \, \mathrm{d}Y = \frac{3}{4} \int_0^\infty e^{-Y^2} \, \mathrm{d}y = \frac{3}{8} \int_{-\infty}^\infty e^{-Y^2} \, \mathrm{d}Y = \frac{3}{8} \pi^{1/2}$$

$$(10.2.11)$$

Insertion into (10.2.10) gives the expected result after some cancelling

$$\overline{s^2} = 3kT/m, \qquad \bar{\epsilon} = \frac{1}{2} m\overline{s^2} = \frac{3}{2} kT \quad (10.2.12)$$

Figure 10.2 shows the relation between the three speeds, the most probable speed, the average or mean speed, and the root mean square speed.

10.3 FRACTION OF MOLECULES IN A GAS WITH TRANSLATIONAL ENERGIES IN EXCESS OF ϵ_0

For many chemical purposes it is necessary to know the fraction of molecules which have translational energies in excess of a certain value, say ϵ_0. Bimolecular chemical reaction for instance occur only when collisions between molecules are sufficiently violent, that is when the velocity resolved along the line of centres of the molecules exceeds a certain value say v_0 at the moment of impact.

We consider first the fraction of molecules with velocities exceeding v_0 in one dimension only. From equation (10.1.2) the fraction is given by (10.3.1). The factor 2 is required to include molecules moving in both directions.

$$f_{1\text{-D}} = 2\left(\frac{m}{2\pi kT}\right)^{1/2} \int_{v_0}^{\infty} \exp\left[-\frac{mv_x^2}{2kT}\right] dv_x \qquad (10.3.1)$$

Making the substitution $Y^2 = mv_x^2/2kT$ or $v_x = (2kT/m)^{1/2} Y$, the right-hand side of 10.3.1 becomes

$$f_{1\text{-D}} = \frac{2}{\sqrt{\pi}} \int_{Y_0}^{\infty} \exp[-Y^2] dY \qquad (10.3.2)$$

This integral cannot be evaluated explicitly, but an approximation for high values of Y can be derived. This is the situation normally of interest in chemistry for it is usually particularly energetic molecules which have a chemically interesting role. The condition $Y_0 \gg 1$ means that the kinetic energy greatly exceeds kT. We proceed by rearranging (10.3.2)

$$f_{1\text{-D}} = \frac{2}{\sqrt{\pi}} \int_{Y_0}^{\infty} \exp[-(Y - Y_0)^2 - 2Y_0(Y - Y_0) - Y_0^2] d(Y - Y_0)$$

$$= \frac{2}{\sqrt{\pi}} \int_{0}^{\infty} \exp[-X^2] \exp[-2Y_0 X] \exp[-Y_0^2] dX \qquad (10.3.3)$$

where $X = Y - Y_0$. We now examine the form of the function inside the integral. Figure 10.3 shows the first and second factors plotted against X for $Y_0 = 5$. It is obvious that the major contribution to the integral arises from the low values of X where the first factor $\exp[-X^2]$ is close to unity. To a good approximation for Y_0 large we can therefore write

$$f_{1\text{-D}} \approx \frac{2}{\sqrt{\pi}} \exp[-Y_0^2] \int_{0}^{\infty} \exp[-2Y_0 X] dX = (\pi Y_0^2)^{-1/2} \exp[-Y_0^2]$$

$$= \left(\frac{2kT}{\pi mv_0^2}\right)^{1/2} \exp\left[-\frac{mv_0^2}{2kT}\right] = \left(\frac{kT}{\pi \epsilon_0}\right)^{1/2} \exp\left[-\frac{\epsilon_0}{kT}\right] \qquad (10.3.4)$$

For motion in two dimensions the result is much simpler. We first

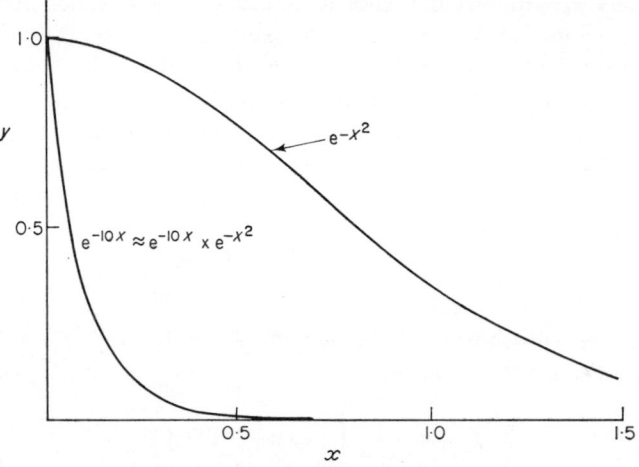

Figure 10.3   Functions in equation (10.3.3).

require the velocity distribution function for two dimensions. By analogy with (10.1.2) the result is

$$p(v_x, v_y)\, dv_x\, dv_y = \frac{m}{2\pi kT} \exp\left[-(v_x{}^2 + v_y{}^2)\frac{m}{2kT}\right] dv_x\, dv_y \quad (10.3.5)$$

To determine the distribution function for speed in two dimensions we proceed by analogy with the three-dimensional case. The speed component of any molecule in the (x, y)-plane is

$$s^2 = v_x{}^2 + v_y{}^2 \tag{10.3.6}$$

This represents a circle of radius s. For all points on the circumference of a circle of radius s the speed is the same, and thus the area element $dv_x\, dv_y$ may be replaced by the area of a thin circular ring of radius s and thickness ds. The area of this ring is $2\pi s\, ds$, whence from (10.3.5)

$$p(s)\, ds = (2\pi s)\frac{m}{2\pi kT} \exp\left[-\frac{s^2 m}{2kT}\right] ds = \frac{ms}{kT} \exp\left[-\frac{ms^2}{2kT}\right] ds \quad (10.3.7)$$

The fraction of molecules with speeds exceeding s_0 is then

$$f_{2\text{-D}} = \int_{s_0}^{\infty} \frac{m}{kT} \exp\left[-\frac{ms^2}{2kT}\right] s\, ds = \int_{X_0}^{\infty} e^{-X}\, dX = e^{-X_0} \quad (10.3.8)$$

where $X = ms^2/2kT$. The fraction is therefore

$$f_{\text{2-D}} = \exp\left[-\frac{ms_0{}^2}{2kT}\right] = \exp\left[-\frac{\epsilon_0}{kT}\right] \qquad (10.3.9)$$

For three-dimensional motion the result is again more complicated and by the method used for one-dimensional motion it may be shown to be

$$f_{\text{3-D}} \approx \left(\frac{\epsilon_0}{2\pi kT}\right)^{1/2} \exp\left[-\frac{\epsilon_0}{kT}\right], \quad \epsilon_0 \gg kT \qquad (10.3.10)$$

Summarizing the three cases we have

$$f_{\text{1-D}} \approx \left(\frac{1}{\pi}\right)^{1/2} \left(\frac{\epsilon_0}{kT}\right)^{-1/2} \exp\left[-\frac{\epsilon_0}{kT}\right], \quad \epsilon_0 \gg kT$$

$$f_{\text{2-D}} = \exp\left[-\frac{\epsilon_0}{kT}\right], \quad \text{all } \epsilon_0$$

$$f_{\text{3-D}} \approx \left(\frac{1}{2\pi}\right)^{1/2} \left(\frac{\epsilon_0}{kT}\right)^{1/2} \exp\left[-\frac{\epsilon_0}{kT}\right], \quad \epsilon_0 \gg kT \qquad (10.3.11)$$

For any real situation in chemistry the exponential term always dominates the pre-exponential term which is of the order of unity. It is thus nearly always an adequate approximation to use the form $f_{\text{2-D}}$ for the fraction of molecules with velocities or speeds in excess of a minimum value.

In particular in connection with bimolecular chemical reactions, it may be shown that the fraction of collisions between spheres in which the velocity, resolved along the line of centres at the moment of impact, exceeds v_0 is exactly $\exp[-\epsilon_0/kT]$. The complete formula for the collision rate between spherical molecules when this condition holds is

$$Z_{12} = \frac{N_1 N_2}{V^2} \pi D_{12}{}^2 \left(\frac{8kT}{\pi\mu}\right)^{1/2} \exp\left[-\frac{\epsilon_0}{kT}\right] \qquad (10.3.12)$$

where (N_1/V) and (N_2/V) are the molecular concentrations, D_{12} is the mean molecular diameter, that is $(D_1 + D_2)/2$, and μ is the reduced mass of the colliding pair, that is $m_1 m_2/(m_1 + m_2)$. In this formula the factor $(8kT/\pi\mu)^{1/2}$ is the average speed of approach of the molecules along the line of centres at the moment of impact. When the molecules are of the same species this formula counts every collision twice and the right-hand side of the equation has to be divided by 2, the symmetry number of the collision complex. For further details on collision theory the reader is referred to Chapter 12 of Fowler and Guggenheim, and to Benson's *Foundations of Chemical Kinetics*.

10.4 THE PRESSURE OF AN IDEAL GAS

The pressure of an ideal gas was evaluated above using the results of classical thermodynamics and molecular thermodynamics combined. In the process of obtaining the expression $pV = NkT$ we justified the identity $\beta = 1/kT$ by comparison of functions which were derivable both from classical thermodynamics and molecular thermodynamics. Although it is mathematically more long-winded, it is conceptually simpler to make the identification using the Maxwell–Boltzmann distribution law and the assumption that the pressure of an ideal gas arises from the averaging of the effects of the impacts of myriads of molecules on the walls of the containing vessel.

We note first of all that equation (10.0.1) arises from the fundamental assumption of molecular thermodynamics, and that in deriving it one does not have to assume the identity $\beta = 1/kT$. In all the succeeding formulae we could therefore have written β for $1/kT$, and in particular we could therefore have written the speed distribution function (10.2.3) in the form:

$$p(s)\,ds = (m\beta/2\pi)^{3/2} \exp\left[-\beta ms^2/2\right] 4\pi s^2\,ds \qquad (10.4.1)$$

We consider the effect of the impacts of molecules on a small area of wall A in a time δt (see Figure 10.4). Initially we focus attention on a small volume of gas dV which is a distance r from A and subtends an angle θ with the normal to A.

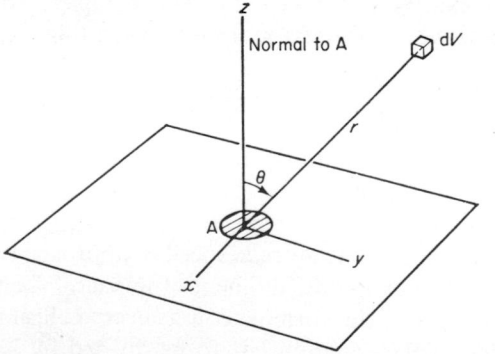

Figure 10.4   Coordinates of a volume element dV relative to an area of surface A.

Only those molecules in dV which are moving in the correct direction will hit A. The fraction of all molecules in dV which can conceivably strike A is obtained from the solid angle subtended by A at the centre of the volume element dV. In steradians this angle is

$$\delta\phi = A \cos \theta/r^2 \qquad (10.4.2)$$

The fraction of molecules which can conceivably strike A is then

$$f = A \cos \theta/4\pi r^2 \qquad (10.4.3)$$

The same is true for all volume elements lying on a ring around the normal as shown in Figure 10.5. The volume element can therefore be

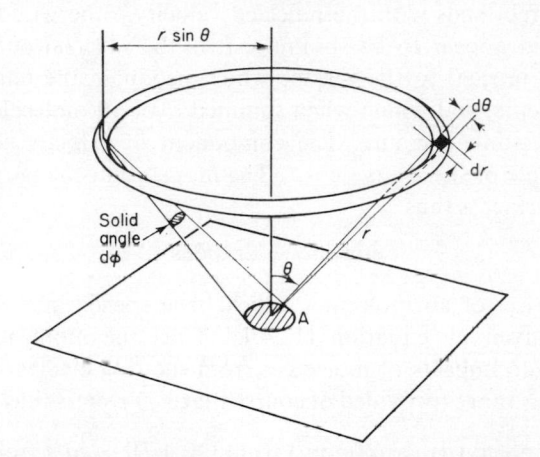

Figure 10.5 Configuration of ring element in relation to an area of surface, A.

expanded to a thin ring. The radius of the ring is $r \sin \theta$ and its cross-sectional area may be written as $r \, dr \, d\theta$. Its volume is therefore

$$dV = (2\pi r \sin \theta) \times (r \, dr \, d\theta) = 2\pi r^2 \sin \theta \, dr \, d\theta \qquad (10.4.4)$$

If the average molecular concentration is $c = N/V$ the number of molecules in the ring element which can conceivably hit A is

$$dN = (\text{concentration}) \times \{(\text{solid angle subtended by } A)/4\pi\} \times$$
$$(\text{volume of ring element})$$

$$= c \times (A \cos \theta/4\pi r^2) \times (2\pi r^2 \sin \theta \, dr \, d\theta)$$

$$= c \times (A/2) \sin \theta \cos \theta \, dr \, d\theta \qquad (10.4.5)$$

Although all these molecules would eventually strike A, not all will strike A within the time δt measured, as it were, after the starting gun. Suppose a molecule has a speed s and its velocity vector points in the correct direction to strike A. The time taken to reach A will be

$$t = r/s \tag{10.4.6}$$

Thus only those molecules whose speeds exceed $r/\delta t$ will strike A within the time δt. When a molecule hits A and rebounds we assume that its velocity normal to the surface is reversed, but its velocity parallel to the surface is unchanged. This will not be the case in practice, but for every molecule which hits the surface with a particular velocity there must be one which rebounds with the reflected velocity, otherwise the gas as a whole would appear to be spinning. It is the reversal of the velocity component normal to the surface which produces the impulse of the single molecule, and which when summed over all molecules gives rise to the observable pressure. The component of velocity normal to A for a molecule of speed s is $s \cos \theta$. The momentum change perpendicular to the surface is thus

$$\delta(\text{mom}) = 2ms \cos \theta \tag{10.4.7}$$

The fraction of all molecules which have speeds in the range s to $s + ds$ is given by equation (10.4.1). Thus the momentum change accruing from impacts of molecules from the ring element which have speeds in this range (provided of course that $s \geqslant r/\delta t$) is thus

$$d(\text{mom})_1 = \{p(s)\, ds\} \times \{\delta(\text{mom}) \text{ from } (10.4.7)\} \times \{dN \text{ from } (10.4.5)\}$$

$$= \left\{ \left(\frac{m\beta}{2\pi}\right)^{3/2} \exp\left[-\frac{\beta ms^2}{2}\right] 4\pi s^2\, ds \right\} \times \{2ms \cos \theta\}$$

$$\times \left\{ \left(\frac{c.A}{2}\right) \sin \theta \cos \theta\, dr\, d\theta \right\} \tag{10.4.8}$$

The contribution to the total pressure is obtained from the change of momentum by dividing first by δt, which gives the force (rate of change of momentum), and then by the area A, which gives the pressure (force per unit area). Thus the contribution to the pressure from molecules in the ring element which have velocities in the range s to $s + ds$ is

$$dp_1 = \left(4\pi c\, \frac{m}{\delta t}\right)\left(\frac{m\beta}{2\pi}\right)^{3/2} \sin \theta \cos^2 \theta\, s^3 \exp\left[-\frac{\beta ms^2}{2}\right] dr\, ds\, d\theta \tag{10.4.9}$$

The total pressure is obtained by integration of dp_1 over all possible values of s, r and θ. The limits of θ are 0 to $\pi/2$ as seen from Figure 10.5. The limits for r and s are interlinked. It turns out to be simplest to integrate first with respect to r, whose limits are $0 < r < s\,\delta t$, and then with respect to s whose limits are $0 < s < \infty$.

Since θ occurs only in the trigonometric functions we integrate with respect to it first. The "θ-integral", I_θ, may be written

$$I_\theta = \int_0^{\pi/2} \cos^2\theta \sin\theta \, d\theta$$

$$= \int_0^{\pi/2} (-\cos^2\theta)\, d\cos\theta$$

$$= -[\cos^\theta \theta/3]_0^{\pi/2} = \tfrac{1}{3} \tag{10.4.10}$$

Inserting this value gives:

$$dp_2 = \left(4\pi c\,\frac{m}{\delta t}\right)\left(\frac{m\beta}{2\pi}\right)^{3/2}\frac{1}{3}\exp\left[-\frac{\beta m s^2}{2}\right] s^3 \, dr \, ds \tag{10.4.11}$$

Integrating now with respect to r we obtain for the "r-integral", I_r

$$I_r = \int_0^{s\,\delta t} dr = s\,\delta t \tag{10.4.12}$$

whence after insertion into (10.4.10) and cancellation of δt gives

$$dp_3 = \left(4\pi c\,\frac{m}{3}\right)\left(\frac{m\beta}{2\pi}\right)^{3/2}\exp\left[-\frac{\beta m s^2}{2}\right] s^4 \, ds \tag{10.4.13}$$

Making the substitution $s = (2/\beta m)^{1/2} Y$, equation (10.4.13) reduces to:

$$dp_3 = \left(4\pi c\,\frac{m}{3}\right)\frac{2}{m\beta}\,\pi^{-3/2}\exp\left[-Y^2\right] Y^4 \, dY \tag{10.4.14}$$

Integration with respect to Y is performed from 0 to ∞, and gives the result quoted in equation (10.2.11), namely $I_Y = \tfrac{3}{8}\pi^{1/2}$.

The final expression for the pressure obtained after this last integration is

$$p = \left(4\pi c\,\frac{m}{3}\right)\frac{2}{m\beta}\,\pi^{-3/2}\frac{3}{8}\pi^{1/2} = \frac{c}{\beta} = \frac{N}{\beta V} \tag{10.4.15}$$

This equation of state may be directly compared with the ideal gas law $p = nRT/V$ where n is the number of moles in the volume V, or

$p = NkT/V$, where N is the number of molecules in the volume, V. We thus obtain the identification

$$\beta = 1/kT \qquad\qquad (10.4.16)$$

This derivation is essentially independent of that made previously in Chapter 4. It does not, however, avoid the necessity of making further connecting assumptions in order to obtain expressions for the entropy, free energy and so on.

PROBLEMS

10.1 Calculate the mean speed of oxygen molecules at 300 K. The speed of sound in a gas is $v = (p\gamma/\rho)^{1/2}$ where $\gamma = C_p/C_v$, ρ = density and p = pressure. Compare v with the molecular velocity.

10.2 In a molecular beam experiment atoms emerge from a furnace and pass through a series of collimating slits which produce a thin beam of atoms all moving in the same direction. What is their mean speed in terms of the furnace temperature and molecular mass? Calculate its numerical value for Cs atoms emerging from a furnace at 1000 K.

10.3 Suppose the collimated beam of Cs atoms described in Problem 10.2 passes through a velocity selector which passes only molecules with velocities between $0.950\, v_{\text{mean}}$ and $1.050\, v_{\text{mean}}$. Roughly what proportion of the original stream will be passed by the selector?

(Insert v_{mean} as obtained in Problem 10.2 into equation (10.1.2) and replace dv_x by $0.100\, v_{\text{mean}}$.)

10.4 What fraction of H_2 and N_2 molecules at the surface of the earth possess velocity components away from the centre in excess of the escape velocity?

The escape velocity is found as follows: If the radius of the earth is R, and its mass M, the force of attraction on any mass m at distance z above the surface is

$$f = GmM/(R + z)^2$$

where $G = 6.67 \times 10^{-11}$ N m^2 kg^{-2} is the gravitational constant. The energy required for escape is then obtained as

$$\epsilon_0 = \int_0^\infty f\, dz$$

Molecules with kinetic energy away from the centre in excess of ϵ_0 can escape. Use equation (10.3.4) to evaluate the necessary fractions of H_2 and

N_2 at 300 K. Carry out the same calculation for the moon. The radii and masses are:

Mass of earth $= 6.0 \times 10^{24}$ kg,   Radius of earth $= 6.4 \times 10^6$ m

Mass of moon $= 7.4 \times 10^{22}$ kg,   Radius of moon $= 1.74 \times 10^6$ m

What do these calculations indicate about the atmospheres of the earth and moon?

10.5   Prove equation (10.3.10).

10.6   Starting from equation (10.0.1) derive an expression for the variation of pressure with height above the earth's surface. Assume that the temperature and acceleration due to gravity are constant.

10.7   Dipole orientation in an electric field. Polar molecules may be regarded as electrical dipoles with a dipole moment $\mu = er$ (see figure)

where e is the charge on each end of the dipole and r the distance between the charges. Prove that the energy of a dipole when orientated at an angle θ to an electric field of strength F is $\epsilon = \mu F(1 - \cos \theta)$. In the absence of a field all orientations in space are equally probable. Hence show that the probability of finding a dipole orientated between θ and $\theta + d\theta$ to any direction is $\frac{1}{2} \sin \theta \, d\theta$. Using an appropriate form of (10.0.1) show that in a field of strength F this probability becomes

$$p(\theta) \, d\theta = \exp [\mu F \cos \theta / kT] \times \tfrac{1}{2} \sin \theta \, d\theta / I$$

where I is the integral of the numerator over the range $0 < \theta < \pi$. Assume that $\mu F / kT \ll 1$ and integrate after expanding to the first term of the exponential series. Now obtain the apparent or mean dipole moment in the direction of the field

$$\bar{\mu} = \int \mu \cos \theta \, p(\theta) \, d\theta$$

Again integrate expanding the exponential to the first term only, and so show that $\bar{\mu} = \mu^2 F / 3kT$. Since $\bar{\mu}$ can be simply related to the dielectric constant of the medium, measurement of the dielectric constant of a gas gives direct information about molecular dipole moments.

Check finally that $\mu F / kT$ is indeed much less than unity. Consider two electronic charges separated by 10^{-10} m as the dipole and a field strength of 10^4 V m^{-1}.

CHAPTER 11

CHEMICAL EQUILIBRIUM

It is widely observed that many chemical reactions, and indeed all reactions if one were to be rigorous, proceed not to completion but to states of equilibrium in which both reactants and products are present in finite concentration. Well known examples of such reactions occurring in the gas phase are:

$$CO + H_2O \rightleftharpoons CO_2 + H_2 \qquad (11.0.1)$$

$$H_2 + I_2 \rightleftharpoons 2HI \qquad (11.0.2)$$

$$(11.0.3)$$

Thermodynamic arguments, kinetic arguments and experiments show that for a general reaction of the type (11.0.4),

$$a.A + b.B + \cdots \rightleftharpoons m.M + n.N + \cdots \qquad (11.0.4)$$

there always exists an equilibrium state which allows the definition of an equilibrium constant (11.0.5)

$$K = ([M]^m[N]^n \ldots)/[A]^a[B]^b \ldots) \qquad (11.0.5)$$

where $[X]$ represents the activity of the substance X in any equilibrium mixture. For assemblies whose systems obey the ideal gas law, $[X]$ may be taken as the concentration or the pressure of the species, X. In this chapter we obtain a molecular thermodynamic expression for K in terms of molecular partition functions which enables K in general to be calculated from the molecular properties of the reactant and product species.

11.1 DERIVATION OF THE MOLECULAR THERMODYNAMIC EQUATION FOR EQUILIBRIUM USING CLASSICAL THERMODYNAMIC EQUATIONS

The classical thermodynamic equation governing the equilibrium (11.0.4) (see equation 3.2.34) is

$$a \cdot \mu_a'^e + b \cdot \mu_b'^e + \cdots = m \cdot \mu_m'^e + n \cdot \mu_n'^e + \cdots \qquad (11.1.1)$$

where a, b, etc. are the stoichiometric numbers, and the $\mu_i'^e$ are the molar chemical potentials of the species in their equilibrium states. The molecular thermodynamic equivalent of (11.1.1) is

$$N_a \cdot \mu_a^e + N_b \cdot \mu_b^e + \cdots = N_m \cdot \mu_m^e + N_n \cdot \mu_n^e + \ldots \qquad (11.1.2)$$

where the μ_i are the chemical potentials per system, and N_a, N_b, etc. are the numbers of molecules of sorts A, B, etc. Since these numbers must be in the same ratio as the numbers of moles reacting, equation (11.1.2) may be rewritten with a, b, etc. replacing N_a, N_b etc:

$$a \cdot \mu_a^e + b \cdot \mu_b^e + \cdots = m \cdot \mu_m^e + n \cdot \mu_n^e + \cdots \qquad (11.1.3)$$

The molecular thermodynamic equations which express the dependence of the chemical potentials upon concentration when the reactants and products behave both individually and when mixed as ideal gases are

$$\mu_i^e = -kT \ln q_i^\circ + kT \ln c_i^e \qquad (11.1.4)$$

where the q_i° are the molecular partition functions per unit volume, and the c_i^e are the concentrations of systems at equilibrium. Inserting (11.1.4) into (11.1.3) and cancelling out kT gives

$$-a \ln q_a^\circ - b \ln q_b^\circ - \cdots + a \ln c_a^e + b \ln c_b^e + \cdots$$
$$= -m \ln q_m^\circ - n \ln q_n^\circ - \cdots + m \ln c_m^e + n \ln c_n^e + \cdots \qquad (11.1.5)$$

Rearranging and removing logarithms gives

$$K_c \equiv \frac{(c_m^e)^m (c_n^e)^n \cdots}{(c_a^e)^a (c_b^e)^b \cdots} = \frac{(q_m^\circ)^m (q_n^\circ)^n \cdots}{(q_a^\circ)^a (q_b^\circ)^b \cdots} \qquad (11.1.6)$$

In practice a modification is required to (11.1.6) for in writing equation (11.1.1) from which (11.1.6) is ultimately derived, it is tacitly assumed that the chemical potentials are measured from a self-consistent zero of energy; in normal thermodynamic usage they would have to be chemical potentials of formation of the reactants and products. In molecular thermodynamic terms equation (11.1.3) requires that the energies of the quantum states used in evaluating the q_i°'s must be measured from a common energy zero, which may, for example, be taken as the energy of the atoms composing the products or reactants in the gaseous state at 0 K. Since it is conventional to measure the energies of quantum states for any system from that of the ground state of the system, an "energy correction factor" must be introduced if conventional partition functions are to be used in the expression for the equilibrium constant. According to Section 6.4, equation (6.4.5), the partition function, q_i°, for which the energies of the quantum states are measured from some arbitrary zero, is related to the conventional partition function, $q_i^{\circ\circ}$, by the equation

$$q_i^\circ = q_i^{\circ\circ} \exp\left[-\epsilon_i^\circ/kT\right] \tag{11.1.7}$$

where ϵ_i° is the energy of the ground state of the molecule above the arbitrary zero.

The equilibrium constant thus takes the form

$$K_c = \frac{(q_m^{\circ\circ})^m (q_n^{\circ\circ})^n \cdots}{(q_a^{\circ\circ})^a (q_b^{\circ\circ})^b \cdots} \times \exp\left[\frac{-(m\epsilon_m^\circ + n\epsilon_n^\circ + \cdots) - (a\epsilon_a^\circ + b\epsilon_b^\circ \cdots)}{kT}\right] \tag{11.1.8}$$

$$K_c = \frac{(q_m^{\circ\circ})^m (q_n^{\circ\circ})^n \cdots}{(q_a^{\circ\circ})^a (q_b^{\circ\circ})^b \cdots} \exp\left[-\frac{\Delta\epsilon_0^\circ}{kT}\right] \tag{11.1.9}$$

where $\Delta\epsilon_0^\circ$ is the energy difference between the ground states of the products and reactants, that is the heat of reaction at absolute zero. Since it is normal to use molar heats of reaction rather than molecular heats of reaction, the exponent is often written $-\Delta E_0^\circ/RT$ rather than $-\Delta\epsilon_0^\circ/kT$. The numerical value of the exponent is, of course, unchanged.

Equation (11.1.9) is the final form of the equation for the equilibrium constant in terms of conventional partition functions. In practice, as seen from the examples given later, the various $q_i^{\circ\circ}$ have to be factorized into their component parts. The quantity $\Delta\epsilon_0^\circ$ may be obtained in a number of ways, the two main methods being spectroscopic (applicable to diatomic and some triatomic molecules) and calorimetric (applicable to nearly all molecules). In evaluating $\Delta\epsilon_0^\circ$ it is important to make the correct allowance for the stoichiometry (see equations 11.1.1 and 11.1.2). Finally we note that K_c will have the units of concentration to some power (possibly zero), and since we have been working throughout in concentration units of molecules per unit volume, $(c = N/V)$, K_c will be in similar units. To convert to molar units Avogadro's number must be introduced.

11.2 DERIVATION OF THE EQUILIBRIUM CONDITION FROM THE FUNDAMENTAL ASSUMPTION ONLY

Equation (11.1.9) may be derived directly from the fundamental assumption without recourse to any thermodynamic arguments. All that is required is a slight extension of the concept of the partition function, and generalization of the idea of a system.

The probability that a system in an ideal gaseous assembly is found in a quantum state, i, is

$$p(\text{state}, i) = \frac{\exp\left[-\epsilon_i/kT\right]}{q} \qquad (11.2.1)$$

where q is the partition function, that is the sum of the numerator over all quantum states accessible to the system.

The probability that a system is found in one-of-a-group of states, which might be labelled say $j, k, \ldots z$, is then

$p(\text{states}, j, k, \ldots, z)$

$$= \frac{\exp\left[-\epsilon_j/kT\right] + \exp\left[\epsilon_k/kT\right] + \cdots + \exp\left[-\epsilon_z/kT\right]}{q} \qquad (11.2.2)$$

The numerator of (11.2.2) may be regarded as the partition function for the specified group of states. We may therefore write in general

$$p(\text{group}) = q_{\text{group}}/q \qquad (11.2.3)$$

Where q_{group} is summed over all possible quantum states which are defined to be in the group, and q is summed over all possible quantum states for the system.

If we are interested in the relative probabilities of finding a system taken at random in two groups a and b which contain no common states we obtain

$$\frac{p(\text{group}, a)}{p(\text{group}, b)} = \frac{q_{\text{group}, a}}{q_{\text{group}, b}} \qquad (11.2.4)$$

We now consider a more specific example. In the terminology of molecular thermodynamics we regard the molecule of 1,2-dichloro-ethylene as a system. There is a large number of quantum states accessible to this type of system. For chemical purposes these states may be conveniently divided into those where the two Cl atoms are cis to the C=C bond, and those states where the two Cl atoms are trans to the C=C bond. When a catalyst is present, say a trace of I atoms, the cis- and trans-states rapidly interconvert, but without a catalyst the two forms can be separated relatively simply. We may therefore conceive of two manifolds of quantum states which are interconvertible under certain circumstances, but distinct under others. Because they can be made interconvertible, and yet are distinguishable we can talk about the probability of finding a system taken at random in either of the two groups. Clearly we can apply equation (11.2.4) directly to give:

$$\frac{p(\text{cis})}{p(\text{trans})} = \frac{q_{\text{cis}}}{q_{\text{trans}}} \tag{11.2.5}$$

We now require an interpretation of $p(\text{cis})$ and $p(\text{trans})$ in terms of the numbers of molecules of the two types. Suppose we have a bag containing 53 black balls, and 147 red balls. The probability of a ball being black when selected at random is simply

$$p(\text{black}) = \frac{(\text{number of black balls})}{(\text{total number})} = \frac{53}{200} \tag{11.2.6}$$

The relative probability of selecting a black relative to a red ball is then

$$\frac{p(\text{black})}{p(\text{red})} = \frac{53}{147} = \frac{(\text{number of black balls})}{(\text{number of red balls})} \tag{11.2.7}$$

The same argument applies to molecules and so we obtain

$$\frac{p(\text{cis})}{p(\text{trans})} = \frac{N_{\text{cis}}}{N_{\text{trans}}} \tag{11.2.8}$$

Whence

$$K_N = \frac{N_{\text{cis}}}{N_{\text{trans}}} = \frac{q_{\text{cis}}}{q_{\text{trans}}} \tag{11.2.9}$$

Dividing numerators and denominators by V we finally obtain

$$K_c = \frac{c_{\text{cis}}}{c_{\text{trans}}} = \frac{q^{\circ}_{\text{cis}}}{q^{\circ}_{\text{trans}}} \tag{11.2.10}$$

Equation (11.2.10) is an elementary version of equation (11.1.6)

The situation is more involved when the equilibrium involves more than two molecular species. Consider for example the equilibrium (11.2.11) involving three atoms, N, O and Cl.

$$NOCl \rightleftharpoons NO + Cl \qquad (11.2.11)$$

To treat this example successfully the basic "system" of the assembly must be defined as a trio-of-atoms, namely (N, O, Cl). Associated with this system of three atoms constrained to a volume V, there is a large number of quantum states which are accessible. They include those states in which the three atoms are independently translating, those in which they are bound together in the form of what we should term an NOCl molecule, and those in which N and O are bound together as NO but the Cl atom free. Other conceivable configurations are NCl + O and OCl + N, ClNO and NClO. The accessibility of these states, and their existence has nothing to do with the concentrations of the species, but results from the principles of quantum mechanics. The complete partition function for the (N, O, Cl) system will contain terms for all possible arrangements of the three atoms. In making up this partition function the energy of every quantum state must be measured from the same base energy, which can conveniently be taken as the energy of the separated atoms N + O + Cl at 0 K in the gas phase.

Proceedings as before using equation (11.2.3) we may write immediately

$$p(NOCl) = \frac{q_{NOCl}}{q_{tot}} \qquad (11.2.12)$$

$$p(NO + Cl) = \frac{q_{(NO+Cl)}}{q_{tot}} \qquad (11.2.13)$$

$$\frac{p(NO + Cl)}{p(NOCl)} = \frac{q_{(NO+Cl)}}{q_{NOCl}} \qquad (11.2.14)$$

The meaning of $q_{(NOCl)}$ is clear enough, but what is to be understood by $q_{(NO+Cl)}$, the partition function of (NO + Cl)? To find the answer we proceed according to the general rules for evaluating partition functions. By definition of the partition function we have

$$q_{(NO+Cl)} = \sum_{states} \exp\left[-\frac{\epsilon_{i\,(NO+Cl)}}{kT}\right] \qquad (11.2.15)$$

where $\epsilon_{i\,(NO+Cl)}$ is the energy of the ith quantum state of the (N, O, Cl) trio in the form NO + Cl. Since the NO and Cl are independent species the energy of any combined quantum state is the sum of the energies of the independent quantum states of NO and Cl. The summation can therefore be split into two factors by the corollary of the multiplication theorem.

$$q_{(NO+Cl)} = \sum_j \exp\left[-\frac{\epsilon_{j(NO)}}{kT}\right] \sum_k \exp\left[-\frac{\epsilon_{k(Cl)}}{kT}\right]$$

$$= q_{NO}q_{Cl} \qquad (11.2.16)$$

Equation (11.2.14) thus becomes

$$\frac{p(NO + Cl)}{p(NOCl)} = \frac{(q_{NO}q_{Cl})}{q_{NOCl}} \qquad (11.2.17)$$

We finally require an interpretation of $p(NOCl)$ and $p(NO + Cl)$ in terms of the numbers of molecules of NOCl, NO and of Cl atoms. By definition $p(NOCl)$ is the probability that an (N, O, Cl) trio taken at random turns out on examination to be a single NOCl molecule. It is thus given by

$$p(NOCl) = \frac{\text{Number of ways of selecting NOCl molecules}}{\begin{cases}\text{Number of ways of selecting (N, O, Cl) trios} \\ \text{irrespective of their mode of combination}\end{cases}} \qquad (11.2.18)$$

Suppose that the number of free Cl atoms is N_{Cl} and the numbers of molecules are N_{NO} and N_{NOCl}. Equation (11.2.18) may then be written

$$p(NOCl) = \frac{N_{NOCl}}{N_{NOCl} + (N_{Cl} \times N_{NO})} \qquad (11.2.19)$$

By the same sort of argument we obtain

$$p(NO + Cl) = \frac{\begin{cases}\text{Number of ways of selecting (N, O, Cl) trios} \\ \text{in the form NO + Cl}\end{cases}}{\begin{cases}\text{Number of ways of selecting (N, O, Cl) trios} \\ \text{irrespective of their mode of combination}\end{cases}}$$

$$= \frac{N_{NO} \times N_{Cl}}{N_{NOCl} + (N_{NO} \times N_{Cl})} \qquad (11.2.20)$$

Substituting (11.2.19) and (11.2.20) into (11.2.17) gives finally

$$K_N \equiv \frac{N_{NO} \times N_{Cl}}{N_{NOCl}} = \frac{q_{NO} \times q_{Cl}}{q_{NOCl}} = \frac{q_{NO}^\circ \times V \times q_{Cl}^\circ \times V}{q_{NOCl}^\circ \times V} \qquad (11.2.21)$$

Transferring the V's to the left-hand side of the equation gives

$$K_c \equiv \frac{c_{NO} \times c_{Cl}}{c_{NOCl}} = \frac{q^\circ_{NO} \times q^\circ_{Cl}}{q^\circ_{NOCl}} \qquad (11.2.22)$$

and employing conventional partition functions gives (11.2.23)

$$K_c = \frac{q^{\circ\circ}_{NO} \times q^{\circ\circ}_{Cl}}{q^{\circ\circ}_{NOCl}} \exp\left[-\frac{\Delta\epsilon^\circ_0}{kT}\right] \qquad (11.2.23)$$

where

$$\Delta\epsilon^\circ_0 = \epsilon^\circ_{NO} + \epsilon^\circ_{Cl} - \epsilon^\circ_{NOCl}. \qquad (11.2.24)$$

Equation (11.2.23) is a version of equation (11.1.9). The argument used for the $NOCl \rightleftharpoons NO + Cl$ equilibrium can be extended to more complex cases without difficulty. Thus the general equation (11.1.9) may be arrived at without reference to classical thermodynamics. Indeed all that is required is the fundamental assumption of molecular thermodynamics, and the identification at some stage of β with $1/kT$. This identification, as we saw in Section 10.4, can also be made without using thermodynamic assumptions.

11.3 NUMERICAL EXAMPLES

Equation (11.1.9) although adequate in itself for all chemical equilibrium applications may usefully be broken down into parts. The unit-volume molecular partition function $q^{\circ\circ}$ may of course be written in terms of the partition functions for different modes of motion

$$q^{\circ\circ} = q^\circ_{trans} q^\circ_{rot} q_{vib} \qquad (11.3.1)$$

It follows that the expression for K_c on the right-hand side of equation (11.1.9) may be similarly split into factors

$$K_c = K_{c\,(trans)} K_{c\,(rot)} K_{c\,(vib)} \exp\left[-\Delta\epsilon^\circ_0/kT\right] \qquad (11.3.2)$$

where $K_{c\,(trans)}$ etc. have the form

$$K_{c(trans)} = \frac{\{q^\circ_{m(trans)}\}^m \{q^\circ_{n(trans)}\}^n \cdots}{\{q^\circ_{a(trans)}\}^a \{q^\circ_{b(trans)}\}^b \cdots} \qquad (11.3.3)$$

The usefulness of this notation is seen when applied in specific examples, for it enables the various component factors in K_c to be evaluated independently.

Example 11.1   The equilibrium constant K_c at 1000 K for the reaction

$$H_2 + D_2 = 2HD$$

This is a straightforward example since nearly all fundamental constants cancel between the numerator and denominator of K_c. We evaluate $K_{c\,(\text{trans})}$, $K_{c\,(\text{rot})}$ and $K_{c\,(\text{vib})}$ separately.

$$K_{c\,(\text{trans})} = \left\{ \frac{(2\pi m_{HD}kT/h^2)^2}{(2\pi m_{H_2}kT/h^2)\,(2\pi m_{D_2}kT/h^2)} \right\}^{3/2}$$

$$= \left\{ \frac{m_{HD}^2}{(m_{H_2}m_{D_2})} \right\}^{3/2} = \left\{ \frac{M_{HD}^2}{(M_{H_2}M_{D_2})} \right\}^{3/2}$$

where M is the molecular weight.

$$K_{c\,(\text{rot})} = \frac{(8\pi^2 I_{HD}kT/h^2)^2}{(8\pi^2 I_{H_2}kT/h^2)\,(8\pi^2 I_{D_2}kT/h^2)} \times \frac{\sigma_{H_2}\sigma_{D_2}}{\sigma_{HD}^2}$$

$$= \frac{4 I_{HD}^2}{(I_{H_2}I_{D_2})} = 4 \left(\frac{\mu_{HD}^2}{\mu_{H_2}\mu_{D_2}} \right) \times \left(\frac{R_{HD}^2}{R_{H_2}R_{D_2}} \right)^2$$

where R is the bond distance and may be taken in Å since the units cancel, and μ is the reduced molecular mass in amu.

$$K_{c\,(\text{vib})} = \frac{(1 - \exp[-\theta_{v,\,H_2}/T])\,(1 - \exp[-\theta_{v,\,D_2}/T])}{(1 - \exp[-\theta_{v,\,HD}/T])^2}$$

The energies of the vibrational levels used in evaluating $K_{c\,(\text{vib})}$ are taken from the ground vibrational state. Thus in evaluating $\Delta\epsilon_0^\circ$ we have to determine the energy difference between these levels not between the classical potential energy zero's for reactants and products.

Table 11.1   Molecular properties of hydrogen isotopes

Property	H_2	D_2	HD
Internuclear distance $10^{10}\,R_e/m$	0·7414	0·7417	0·7413
Molecular weight/amu	2·015	4·028	3·022
Reduced molecular mass/amu	0·5039	1·0070	$\dfrac{1\cdot008 \times 2\cdot014}{3\cdot022}$
Vibration frequency ω/cm^{-1}	4405	3119	3817
Dissociation energy D_0°/ev [a]	4·4763	4·5536	4·5112
q_{vib} at 1000 K	1·000	1·003	1·001

[a] 1 ev $\equiv$ 1 electron volt = 96·48 kJ mol^{-1}.

The relevant numerical·data for the three hydrogen isotopic species are given in Table 11.1. The heat reaction at 0 K is

$$\Delta\epsilon_0^\circ \times N_A = \Delta E_0^\circ = -2D_{0(HD)}^\circ + D_{0(H_2)}^\circ + D_{0(D_2)}^\circ$$

$$= (-2 \times 4{\cdot}5112 + 4{\cdot}4763 + 4{\cdot}5536) \times 96{\cdot}48 \text{ kJ mol}^{-1}$$

$$= +0{\cdot}0075 \times 96{\cdot}48 \text{ kJ mol}^{-1}$$

$$= +0{\cdot}723 \text{ kJ mol}^{-1} = 723 \text{ J mol}^{-1}$$

The final expression for K_c is then

$$K_c = \left\{\frac{M_{HD}^2}{M_{H_2}M_{D_2}}\right\}^{3/2} \times \left\{\frac{\mu_{HD}^2}{\mu_{H_2}\mu_{D_2}}\right\} \times 4 \times \left\{\frac{R_{HD}^2}{R_{H_2}R_{D_2}}\right\}^2 \times K_{c\,(vib)}$$

$$\times \exp\left[-\frac{\Delta\epsilon_0^\circ}{kT}\right]$$

$$= \left\{\frac{3{\cdot}022^2}{2{\cdot}015 \times 4{\cdot}028}\right\}^{3/2} \left\{\frac{1{\cdot}008^2 \times 2{\cdot}014^2}{3{\cdot}022^2 \times 0{\cdot}5039 \times 1{\cdot}0070}\right\} \times 4$$

$$\times \left\{\frac{0{\cdot}7413^2}{0{\cdot}7414 \times 0{\cdot}7417}\right\}^2 \times \left\{\frac{1{\cdot}001^2}{1{\cdot}003 \times 1{\cdot}000}\right\}$$

$$\times \exp\left[-\frac{723}{1000 \times 8{\cdot}314}\right]$$

$$= 1{\cdot}194 \times 3{\cdot}557 \times 0{\cdot}999 \times 0{\cdot}999 \times 0{\cdot}917$$

$$= 3{\cdot}887$$

This value of K_c is in excellent agreement with the experimental value of $3{\cdot}8 \pm 0{\cdot}1$, and is close to the value of 4 which would be predicted on the basis of symmetry alone. The effects of the differences in masses and moments of inertia are seen to be small and partially to cancel out. The difference in the zero point energies of the three isotopic species is so small as to have little effect upon equilibrium. The major factor in the final result is the symmetry number.

This example illustrates a further important result.

When the equation for the reaction contains the same numbers of molecules on both sides of the equation, and particularly when these molecules possess in total the same numbers of rotational degrees of freedom, all fundamental constants in $K_{c\,(trans)}$ and $K_{c\,(rot)}$ cancel out. Much computation is thus avoided by carrying out this cancellation at an early stage of the operation.

Example 11.2 *The water–gas reaction*

$$CO_2 + H_2 \rightleftharpoons H_2O + CO$$

Experiment shows that the equilibrium constant for this reaction is 2·0 at 1295 K, 2·8 at 1565 K and 3·5 at 1823 K. We calculate the value at 1565 K by molecular thermodynamics and compare it with the experimental value. The equilibrium constant is dimensionless, and thus all fundamental constants cancel in $K_{c(\text{trans})}$. Unfortunately they do not in $K_{c\,(\text{rot})}$ since the species on the left-hand side of the equation possess four rotational degrees of freedom, while those on the right possess five.

The basic equation for the molecular thermodynamic calculation is:

$$K_c = \frac{q_{H_2O}^{\circ\circ} \cdot q_{CO}^{\circ\circ}}{q_{CO_2}^{\circ\circ} \cdot q_{H_2}^{\circ\circ}} \exp\left[-\frac{\Delta E_0^\circ}{RT}\right]$$

where ΔE_0° is the difference in energy between the ground rotational/vibrational states of the reactants and products.

Since the numbers of molecules on the two sides of the equation for the reaction are the same we immediately obtain

$$K_{c\,(\text{trans})} = \{M_{H_2O} M_{CO}/(M_{CO_2} M_{H_2})\}^{3/2}$$

For rotational factor in K_c we obtain

$$K_{c\,(\text{rot})} = \frac{\sqrt{\pi}(ABC_{H_2O})^{1/2}\left\{\dfrac{8\pi^2 kT}{h^2}\right\}^{3/2} \times I_{CO}\left\{\dfrac{8\pi^2 kT}{h^2}\right\}}{I_{CO_2}\left\{\dfrac{8\pi^2 kT}{h^2}\right\} \times I_{H_2}\left\{\dfrac{8\pi^2 kT}{h^2}\right\}} \times \frac{\sigma_{CO_2}\,\sigma_{H_2}}{\sigma_{H_2O}\,\sigma_{CO}}$$

$$= \left\{\frac{(ABC_{H_2O})^{1/2}\,I_{CO}}{I_{CO_2} I_{H_2}}\right\} \times \left\{\frac{8\pi^3 kT}{h^2}\right\}^{1/2} \times 2$$

If all moments of inertia are expressed in amu Å² a change is necessary in the factor containing the fundamental constants

$$K_{c\,(\text{rot})} = \left\{\frac{2(ABC_{H_2O})^{1/2}\,I_{CO}}{I_{CO_2} I_{H_2}}\right\}\left\{\frac{8\pi^3 kT \times 10^{-20}}{h^2 L_A}\right\}^{1/2} \quad (I\text{'s in amu Å}^2)$$

$$\frac{8\pi^3 kT \times 10^{-20}}{h^2 L_A} = \frac{8 \times \pi^3 \times 1\cdot3804 \times 10^{-23} \times 10^{-20} \times 1\cdot565 \times 10^3}{6\cdot626^2 \times 10^{-68} \times 6\cdot023 \times 10^{26}}$$

$$= 2\cdot034 \times 10^2$$

The vibrational factor in K_c is

$$K_{c\,(\text{vib})} = \frac{q_{\text{vib (H}_2\text{O, 3 modes)}}\, q_{\text{vib (CO, 1 mode)}}}{q_{\text{vib (CO}_2,\, 4\,\text{modes)}}\, q_{\text{vib (H}_2,\, 1\,\text{mode)}}}$$

The molecular and other parameters necessary to obtain K_c are listed in Table 11.2. We require first the moment of inertia product

Table 11.2 Molecular and other parameters for the water–gas reaction

Property	CO$_2$	H$_2$	H$_2$O	CO
Heat of formation at 0 K				
$\Delta E^\circ_{of}/\text{kJ mol}^{-1}$	$-393 \cdot 165$	$0 \cdot 000$	$-238 \cdot 935$	$-113 \cdot 813$
Molecular weight/amu	$44 \cdot 01$	$2 \cdot 016$	$18 \cdot 02$	$28 \cdot 01$
Bond length $10^{10} \times R_e$/m	$1 \cdot 160$	$0 \cdot 741$	$0 \cdot 960$	$1 \cdot 128$
Bond angle/degrees	180		$104 \cdot 5$	
Moment of inertia/(amu Å^2)	$43 \cdot 0$	$0 \cdot 277$	$[1 \cdot 27]^a$	$8 \cdot 73$
Vibration frequencies				
ω/cm$^{-1\,b}$	2350	4160	3652	2168
	1320		3756	
	668		1595	
	668			

^{a} The value for water is ABC. See text.

^{b} The frequencies listed are those for the transitions from the ground to first excited states. They are slightly lower than the fundamental frequencies because of the correction for anharmonicity. This partial correction for anharmonicity gives a somewhat improved accuracy in the calculation of thermodynamic properties.

for the water molecule. Figure 11.1 shows the geometry of the molecule. The centre of mass cannot be found by inspection, but it is clear that the principal axes of rotation will be the y-axis, a line parallel to the x-axis (which is drawn through the oxygen nucleus), and a line parallel to the z-axis. Taking the O-atom as the origin the coordinates of the two H atoms are

$$y = 0 \cdot 96 \cos 52 \cdot 25^\circ = 0 \cdot 586 \text{ Å}$$

$$x = \pm 0 \cdot 96 \sin 52 \cdot 25^\circ = 0 \cdot 758 \text{ Å}$$

Applying formula (6.2.22) gives

$$\sum m_i x_i^2 = 2 \times 1 \cdot 008 \times 0 \cdot 758^2 = 1 \cdot 16$$

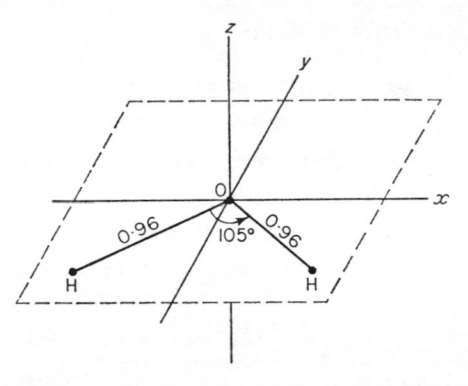

Figure 11.1　Configuration of the water molecule

$$\sum m_i y_i{}^2 = 2 \times 1{\cdot}008 \times 0{\cdot}586^2 - 2(1{\cdot}008 \times 0{\cdot}586)^2/18{\cdot}02$$
$$= 0{\cdot}691 - 0{\cdot}075 = 0{\cdot}616$$
$$\sum m_i z_i{}^2 = 0$$

Whence

$$I_{xx} = 0{\cdot}616, \quad I_{yy} = 1{\cdot}16, \qquad I_{zz} = 1{\cdot}78$$
$$ABC = I_{xx}I_{yy}I_{zz} = 0{\cdot}616 \times 1{\cdot}16 \times 1{\cdot}78$$
$$= 1{\cdot}27 \text{ amu}^3 \text{ Å}^6$$

The vibrational partition functions, the energies of the states being measured from the ground state, are listed in Table 11.3. From them we obtain

$$K_{c\,(\text{vib})} = 1{\cdot}40 \times 1{\cdot}16/(7{\cdot}68 \times 1{\cdot}02) = 0{\cdot}208$$

Table 11.3　Vibrational partition functions for 1565 K

Molecule	ω/cm^{-1}	q_{vib} at 1565 K	q_{vib} total
CO_2	2350	1·13	
	1320	1·43	
	668	2·18	
	668	2·18	7·68
H_2	4160	1·02	1·02
H_2O	3652	1·04	
	3756	1·04	
	1595	1·30	1·40
CO	2168	1·16	1·16

The heat of the reaction at 0 K is

$$\Delta E_0^\circ = -113 \cdot 813 - 238 \cdot 935 + 393 \cdot 165 + 0 \cdot 000$$

$$= +40 \cdot 417 \text{ kJ mol}^{-1}$$

$$\Delta E_0^\circ / RT = 40 \cdot 417 / (1 \cdot 565 \times 8 \cdot 314) = 3 \cdot 115$$

We finally obtain for K_c:

$$K_c = \left\{ \frac{18 \cdot 02 \times 28 \cdot 01}{44 \cdot 01 \times 2 \cdot 016} \right\}^{3/2} \times \left\{ \frac{2 \times \sqrt{1 \cdot 27} \times 8 \cdot 73}{43 \cdot 0 \times 0 \cdot 277} \right\}$$

$$\times (2 \cdot 034 \times 10^2)^{1/2} \times 0 \cdot 208 \times \exp[-3 \cdot 115]$$

$$= 13 \cdot 54 \times 1 \cdot 65 \times 14 \cdot 30 \times 0 \cdot 208 \times 4 \cdot 48 \times 10^{-2}$$

$$= 2 \cdot 98$$

Experimental value $K_c = 2 \cdot 8$.

The calculated and experimental values agree excellently.

Example 11.3 *The heat of dissociation of fluorine.*

The equilibrium constant for this reaction was measured by Doescher (*J. Chem. Physics*, **20**, 330 (1952)), and from the dependence of K_p upon temperature a rough value of ΔH for the dissociation was obtained. A selection of the experimental data is given in Table 11.4. A more

Table 11.4 Equilibrium constant for $F_2 \rightleftharpoons F + F$[a]

Temperature T/K	1115	1010	901	850	760
Equilibrium constant $K_{p(exp)}$/atm	$7 \cdot 55 \times 10^{-2}$	$1 \cdot 20 \times 10^{-2}$	$1 \cdot 29 \times 10^{-3}$	$4 \cdot 12 \times 10^{-4}$	$2 \cdot 0 \times 10^{-5}$

Best Van't Hoff equation through data:

$$\text{Log}_{10}(K_p/\text{atm}) = 14 \cdot 5 - 19\,000(\text{K}/T)$$

Heat of reaction at 1000 K from d ln K_p/dT

$$\Delta H(1000 \text{ K}) = 158 \pm 8 \text{ kJ mol}^{-1}$$

[a] Doescher, *J. Chem. Physics*, **20**, 330 (1952).

accurate value for ΔH was obtained by Stamper and Barrow (*Trans. Faraday Soc.*, **54**, 1592 (1958)) using the methods of molecular thermodynamics. We outline their method here as an example of the general method for determining the heat of a reaction from knowledge of the equilibrium constant at a single temperature.

According to molecular thermodynamics the equilibrium constant for the reaction $F_2 \rightleftharpoons 2F$ is given by

$$K_c = \{(q_F^{\circ\circ})^2/q_{F_2}^{\circ\circ}\} \exp\left[-\Delta E_0^{\circ}/RT\right]$$

The units of K_c will be m^{-3}. To convert to atm. units we use the conversion:

$$\frac{K_p}{atm} = \frac{K_c}{molecule\ m^{-3}} \times \frac{(kT/J)}{1\cdot01325 \times 10^5}$$

To evaluate ΔE_0° we first determine the theoretical value for the ratio $(q_F^{\circ\circ})^2/q_{F_2}^{\circ\circ}$ which will be called $K_{c(q)}$. This is then converted to the equivalent in atm. units, namely $K_{p(q)}$. The ratio $K_{p(q)}/K_{p(exp)}$ should then be $\exp\left[+\Delta E_0^{\circ}/RT\right]$ from which ΔE_0° is readily found. To evaluate $K_{c(q)}$ we require the molecular parameters for F and F_2.

F-atom

The ground electronic state is a ^2P state which is split into a $^2P_{3/2}$ state of degeneracy 4, and a $^2P_{1/2}$ state of degeneracy 2 whose energy is 404 cm^{-1} higher. The electronic partition function for the F atom is thus

$$q_{el} = 4 + 2\exp\left[-404hc/kT\right]$$
$$= 2\{2 + \exp\left[-581(K/T)\right]\}$$

F_2 molecule

The internuclear distance is $1\cdot418$ Å, the fundamental vibration frequency is 892 cm^{-1}.

The moment of inertia of F_2 is then

$$I_{F_2} = 19\cdot00 \times 1\cdot418^2/2 = 19\cdot10\ amu\ Å^2$$

The vibrational partition function is

$$q_{vib,\ F_2} = (1 - \exp\left[-1279(K/T)\right])^{-1}$$

We can now obtain $K_{p(q)}$ as:

$$K_{p(q)} = \frac{kT}{1\cdot0133 \times 10^5} \times \frac{\left(\dfrac{2\pi m_F kT}{h^2}\right)^3}{\left(\dfrac{2\pi m_{F_2} kT}{h^2}\right)^{3/2}} \times \frac{\sigma_{F_2}}{\left(\dfrac{8\pi^2 I_{F_2} kT}{h^2}\right)} \times \frac{(q_{el,\ F})^2}{q_{vib,\ F_2}}$$

Replacing molecular masses, m, by M/L_A, where M is the molecular weight, inserting I in amu $Å^2$, and gathering together all constants we obtain

$$K_{p(q)} = \frac{2^{3/2}k^{3/2} \times 10^{20}}{\pi^{1/2}hL_A^{1/2} \times 1{\cdot}0133 \times 10^5} \times \frac{M_F^3}{M_{F_2}^{3/2}I_{F_2}} \times T^{3/2}$$

$$\times (1 - \exp[-1279/T])(2 + \exp[-581/T])^2$$

$$= 4{\cdot}967 \times 1{\cdot}5328 \times T^{3/2} \times X \times Y^2 = 7{\cdot}613T^{3/2} \times X \times Y^2$$

where $X = (1 - \exp[-1279/T])$ and $Y = (2 + \exp[-581/T])$.

The parameters $T^{3/2}$, X and Y are tabulated in Table 11.5 for the five temperatures listed in Table 11.4. From them five values of $K_{p(q)}$ are

Table 11.5 Data for calculation of ΔE_0° for $F_2 \rightleftharpoons F + F$

T/K	$(T/K)^{3/2}$	X	Y	$K_{p(q)}$/atm
1115	$3{\cdot}723 \times 10^4$	$0{\cdot}6824$	$2{\cdot}5948$	$13{\cdot}022 \times 10^5$
1010	$3{\cdot}210 \times 10^4$	$0{\cdot}7181$	$2{\cdot}5635$	$11{\cdot}534 \times 10^5$
901	$2{\cdot}705 \times 10^4$	$0{\cdot}7582$	$2{\cdot}5257$	$9{\cdot}963 \times 10^5$
850	$2{\cdot}478 \times 10^4$	$0{\cdot}7779$	$2{\cdot}5058$	$9{\cdot}212 \times 10^5$
760	$2{\cdot}095 \times 10^4$	$0{\cdot}8142$	$2{\cdot}4666$	$7{\cdot}901 \times 10^5$

T/K	$K_{p(\text{exp})}$	$K_{p(q)}$	$\ln \dfrac{K_{p(q)}}{K_{p(\text{exp})}}$	$\Delta E_0^\circ = $ $RT\ln \dfrac{K_{p(q)}}{K_{p(\text{exp})}}$
1115	$7{\cdot}55 \times 10^{-2}$	$13{\cdot}022 \times 10^5$	$16{\cdot}663$	$154{\cdot}48$ kJ mol^{-1}
1010	$1{\cdot}20 \times 10^{-2}$	$11{\cdot}534 \times 10^5$	$18{\cdot}383$	$154{\cdot}36$
901	$1{\cdot}29 \times 10^{-3}$	$9{\cdot}963 \times 10^5$	$20{\cdot}465$	$153{\cdot}30$
850	$4{\cdot}12 \times 10^{-4}$	$9{\cdot}212 \times 10^5$	$21{\cdot}528$	$152{\cdot}14$
760	$2{\cdot}00 \times 10^{-5}$	$7{\cdot}901 \times 10^5$	$24{\cdot}400$	$154{\cdot}17$
			Mean	$153{\cdot}68$ kJ mol^{-1}

obtained and hence, by comparison with the experimental values, ΔE_0°. The values over the temperature range 760 to 1115 K are highly consistent, and the mean value of $153{\cdot}68$ is close to the best value obtained by Stamper and Barrow by consideration of all the experimental data of Doescher, namely $153{\cdot}59$ kJ mol^{-1}. Stamper and Barrow claim that the probable error in their mean value is about $0{\cdot}5$ kJ mol^{-1}, a sixteenfold improvement in accuracy over the method based upon the change in $K_{p(exp)}$ with temperature. The reason for the apparent inaccuracy of the van't Hoff method is not the inaccuracy of the data which is the same for both methods, but the restricted temperature range available. By use of the third law the temperature range is effectively extended to 0 K.

PROBLEMS

11.1 In Section 5.2 after equation (5.2.18) it is shown that the approximation $\ln N! = N \ln N - N$ is sufficiently accurate for all applications which involve $\ln q$. In (11.1.6) we derive an expression for K_c in terms of q's not $\ln q$'s. This expression could be in error through the inaccuracy of the approximation used for $N!$ Obtain a more accurate expression for the chemical potential from the equation for F using the very accurate form of Stirling's approximation $\ln N! = N \ln N - N + \frac{1}{2} \ln (2\pi N)$, and thence an accurate expression for K_c. Thus show that the error in (11.1.6) is indeed negligible in practice.

11.2 Evaluate the equilibrium constant K_c at 1000 K for the reaction

$$Cl + H_2 = HCl + H$$

The rotational and vibrational constants for H_2 and HCl are found in Table 6.2. The dissociation energies of H_2 and HCl from their ground vibrational states are $D_0^{\circ}(H_2) = 4\cdot476$ ev, $D_0^{\circ}(HCl) = 4\cdot425$ ev. The ground state of Cl is fourfold degenerate, and the first excited state, 881 cm^{-1} higher in energy, is doubly degenerate. The ground state of H is doubly degenerate.

11.3 The equilibrium constant for the reaction

$$I_2(g) \rightleftharpoons 2I$$

is 0·168 atm at 1275 K and 0·0480 at 1173 K. The vibrational and rotational properties of I_2 are given in Table 6.2. The ground state of I is fourfold degenerate, and the first excited state doubly degenerate at an energy 7603 cm^{-1} higher. Calculate the dissociation energy of I_2 at 0 K. The accepted value is 1·544 ev.

11.4 Calculate the equilibrium constant for dissociation of cyanogen, $C_2N_2 \rightleftharpoons 2CN$ at 2500 K from the following data:

Heat of reaction at 0 K $= \Delta E_0^{\circ} = 475$ kJ mol^{-1},

CN radical: bond distance $= 1\cdot172$ Å; fundamental frequency $= 2062$ cm^{-1}; electronic degeneracy $= 2$.

C_2N_2 molecule: $C\equiv C$ bond distance $= 1\cdot380$ Å; C—N bond distance $= 1\cdot157$ Å. Fundamental frequencies and degeneracies (in brackets) 226 (2), 506 (2), 848, 2149, 2322 cm^{-1} (molecule linear).

The value of K_p calculated by Ruther, McLean and Scheller (*J. Chem. Physics*, **24**, 173 (1956)) for 2500 K is $1\cdot08 \times 10^{-2}$ atm. There is no experimental value for K_p. This is unfortunate since ΔE_0° given above is subject to considerable uncertainty. If K_p were accurately known at a single temperature the dissociation energy could be accurately found.

CHAPTER 12

CHEMICAL KINETICS: THE TRANSITION STATE THEORY

Most overall chemical reactions are complex, and proceed by a sequence of elementary steps, each of which has a definable transition state. Typical of such elementary reactions are:

$$H + Cl_2 \longrightarrow H\text{---}Cl\text{---}Cl \longrightarrow HCl + Cl \qquad (12.0.1)$$

$$CH_3 + CH_3 \longrightarrow H_3C\text{---}CH_3 \longrightarrow C_2H_6 \qquad (12.0.2)$$

$$C_2H_5I \longrightarrow \underset{\overset{|}{H}\text{---}\overset{|}{I}}{H_2C\text{===}CH_2} \longrightarrow C_2H_4 + HI \qquad (12.0.3)$$

$$\text{Cyclo } C_3H_6 \longrightarrow \qquad \longrightarrow \qquad (12.0.4)$$

Elementary reactions (12.0.1) and (12.0.2) are termed "bimolecular" since two molecules of reactant combine to form the transition state, while reactions (12.0.3) and (12.0.4) are termed "unimolecular" since the transition states are formed by the distortion of single molecules of reactant. It is worth noting that the reverse of (12.0.2) is a unimolecular dissociation, and that the reverse of (12.0.3) is a bimolecular association. By the principles of thermodynamics, if the rate constant for any forward reaction and the equilibrium constant are known, then the rate constant for the reverse is determined. This application of the so-called "principle of detailed balancing" is often most valuable in obtaining kinetic information about unknown reactions.

The symbols in equations (12.0.1) to (12.0.4) simultaneously express three quite distinct ideas about the reactions:

(1) They give the overall stoichiometry
(2) They may be taken to represent individual molecules, and there-

216

fore to give a rather crude idea of the geometrical changes which must occur in the formation of the transition state, and its rearrangement to products.

(3) At the molecular thermodynamic level the symbols in the equations imply that the reactants, products and complexes are individually in thermal equilibrium. (Not, of course, that the products are in chemical equilibrium with the reactants.)

It is well known that some elementary reactions (unimolecular rearrangements, dissociations and bimolecular associations) are themselves complex in that their kinetics may be explained only by invoking more fundamental processes called basic reactions. There are four types of basic reaction:

(1) Bimolecular group exchange reactions, which are almost identical to the corresponding elementary reactions; reaction (12.0.1) is an example.

(2) Bimolecular energy exchange reactions. These are activation and deactivation processes which produce or remove vibrationally and rotationally excited molecules by collision.

(3) Bimolecular associations forming excited molecules, and the reverse processes, unimolecular dissociations of excited molecules.

(4) Unimolecular rearrangements of excited molecules.

In the equations for these types of reaction the symbols no longer represent species which are necessarily in thermal equilibrium with their surroundings. It is only for the four types of basic reaction that expressions for the rates can be directly derived using theory. However in the initial treatment of rate theory we shall avoid the complications which arise when basic reactions of excited molecules are important, by treating bimolecular group exchange reactions. The theory of the other types of basic reaction is covered in books on reaction kinetics, for example *Unimolecular Reactions* by Holbrooke and Robinson, Wiley, 1971.

The theory now outlined is variously called the transition state theory, activated complex theory, the theory of absolute reaction rates and it is associated with the names of Eyring and co-workers who made major developments between 1930 and 1940. The authoritative treatment of this theory is given in the classic monograph *Theory of Rate Processes*, by Glasstone, Laidler and Eyring, McGraw-Hill Book Co., 1940. The treatment given below differs in some details from theirs but achieves the same final result. It bears some similarity to that given by Slater in his book *Unimolecular Reactions*.

The progress of any reaction involving three atoms, for example reaction (12.0.1), is most conveniently considered in conjunction with a potential energy diagram (PE diagram) in which the potential energy of the triatomic system is given as a function of the positions of the three atoms.

Three coordinates are required to specify the internal arrangements of the atoms of any triatomic system. These may be taken as the distances r_1, r_2 and θ shown in Figure 12.1. For every combination of r_1,

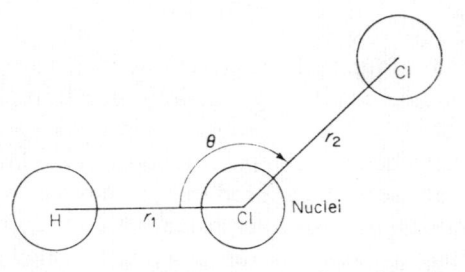

Figure 12.1   Internal coordinates of a triatomic system.

r_2 and θ, the triatomic system H---Cl---Cl has a definite potential energy, which, in principle if not in practice, can be calculated by the methods of quantum mechanics. We can thus assign an energy to every accessible point in the relevant three-dimensional configurational space. If a fourth dimension is used to represent the energy we obtain a three-dimensional PE surface in four dimensional space. But since four-dimensional space is difficult to visualize, we restrict attention initially to one particular value of θ, say $\theta = 180°$. In this way the problem of representation is reduced temporarily to one in three dimensions. It is important to emphasize that we are doing this as a convenience to enable us to visualize the development of the reaction, not because the theory is restricted to a three-dimensional PE diagram.

The PE diagram now becomes a two-dimensional surface in three-dimensional space. It is usual to present the diagram as a map showing contours of equal energy as shown in Figure 12.2. The PE diagram is seen to consist of two valleys, one pointing north–south, and the other

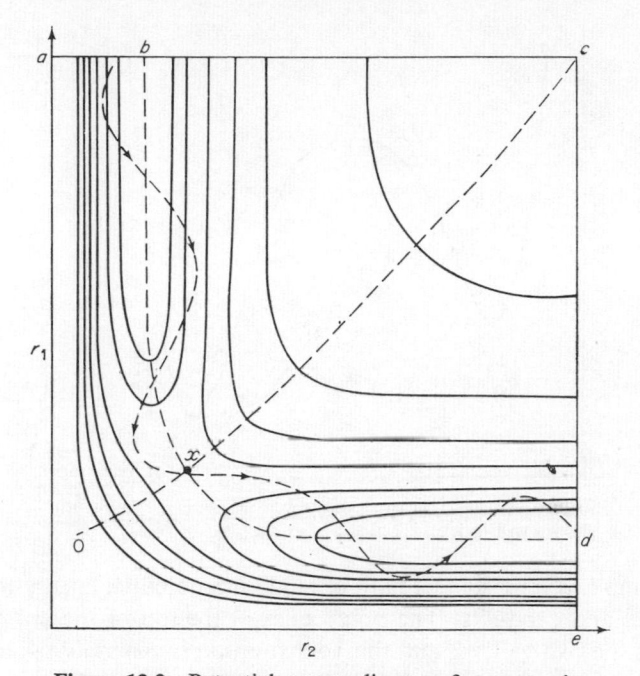

Figure 12.2 Potential energy diagram for a reaction such as $H + Cl_2 = HCl + Cl$, involving three atoms, and requiring only two internal coordinates, r_1 and r_2.

east–west, which are connected by a pass. On either side the valleys rise to highlands. Towards the north–east there is a high plateau, while to the west and south the walls of the valleys rise without limit. The point b represents reactants in their ground state, for it implies that the H—Cl distance, r_1, is large, while the Cl—Cl distance, r_2, is small and furthermore is chosen so that the potential energy is a minimum. Chemically this state is written $H + Cl_2$. Similarly the point d represents the ground state of the products, $HCl + Cl$. A section of the diagram along the line abc will cut the surface to give a potential energy line for the variation of the potential energy when r_2 is varied while r_1 is fixed. This will have the form of the Morse curve for Cl_2. Likewise the section edc will have the form of the Morse curve for HCl. The point c represents the state of complete dissociation of the H—Cl—Cl system into atoms. The region around c is flat because the energy of the three atoms is independent of their positions as long as they are far enough apart. The two Morse curves are shown schematically in Figure 12.3.

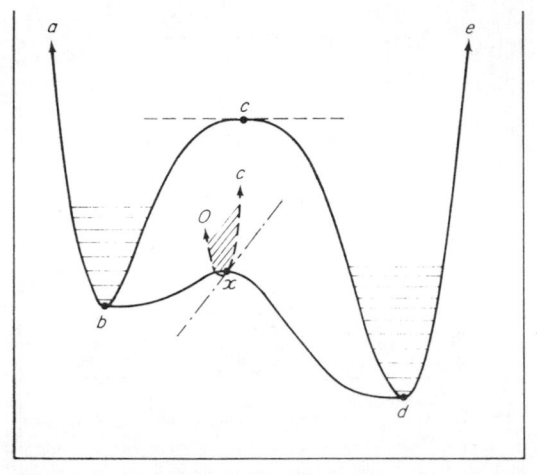

Figure 12.3 Sections of potential energy diagram shown in Figure 12.2 along lines *abc, cde* and *bxd*.

The line *bxd* represents a path of minimum potential energy between reactants and products. The point *x* lies at the top of the col or pass between the two valleys, and this configuration is generally taken to be that of the transition state. It is in fact only the most popular configuration for the transition state, not the only one, as we shall see later. The energy profile along the line *bxd* is shown in Figure 12.3 which illustrates how much easier it is to pass to products via the col than over the high plateau which would involve complete dissociation into atoms.

During reaction the configuration of the trio H—Cl—Cl continually changes, but at any instant is represented by a point which lies somewhere on the PE surface. The development of the reaction of any trio is then shown by the movement of this representative point over the surface. The way in which the point moves is determined by the laws of quantum and classical mechanics.

Suppose we initiate a reaction by firing an H-atom towards a Cl_2 molecule which is stationary along the line of the nuclei. The distance r_1 will rapidly decrease while r_2 will remain more or less constant. The representative point will thus move south along the north–south valley at a speed equal to the translational speed of the H-atom. As the H-atom approaches the Cl_2 molecule it begins to suffer a repulsion and so slows down. The potential energy of the trio increases as the relative translational energy of the pair is converted into potential energy. The representative point begins to climb up the head wall of the valley towards the col. If the initial kinetic energy of the H-atom is too small the H-atom simply bounces off the Cl_2 molecule, and the representative point fails to reach the col; it approaches the col and then retreats, usually by a different route. However, if the initial kinetic energy is sufficient, the representative point passes over the col, and emerges along the east–west valley travelling east. The distance r_2 then increases rapidly while r_1 becomes small and more or less constant. This means that the Cl atom is now moving quickly away from the newly formed HCl molecule. The rate of increase of r_2 gives the translational velocity of the Cl atom relative to the HCl molecule. Reaction is thus represented by passage of the representative point over the col.

Suppose now that the Cl_2 molecule is not initially at rest but is vibrating. This means that the distance r_2 is changing in an oscillatory way. When the H-atom is fired at the Cl_2 molecule the representative point will move off down the valley swinging from side to side of the valley. When it emerges from the reaction it will in all probability be swinging from side to side of the east–west valley, and the HCl product molecule will be vibrationally excited. Such a trajectory is shown by the zig-zag line in Figure 12.2. The passage over the col may thus be at an angle to the line bxd, and obviously need not be precisely at x. Evidently there are many passages over the col, and elsewhere over the energy barrier. We must therefore define "reaction" in a more general way than simply as "passage over the col at x". It is best defined as occuring when the representative point passes over a critical line oxc which is drawn so as to be perpendicular at all points to the energy contours, and to pass through the highest point at the col. We then define "transition states" as those configurations of the activated complex whose representative points lie on this critical line.

Let us now consider what happens if we reinstate θ as a variable. We then have a potential energy diagram such as shown in Figure 12.2 for every possible value of θ. The representative point for a reaction can then move from one such diagram to the next as θ changes simultaneously with r_1 and r_2. Successful reaction may be said to occur whenever the representative point crosses a critical line such as oxc in any of the diagrams. Instead of using a large number of PE diagrams for different values of θ, we could equally well use a four-dimensional representation. The "col" would then be a series of points which would comprise a line. The group of critical lines would form a two-dimensional surface. This critical surface would cut perpendicularly the contours of constant energy, which themselves would be two-dimensional surfaces. This is illustrated in Figure 12.4 which shows the line of the col, the critical

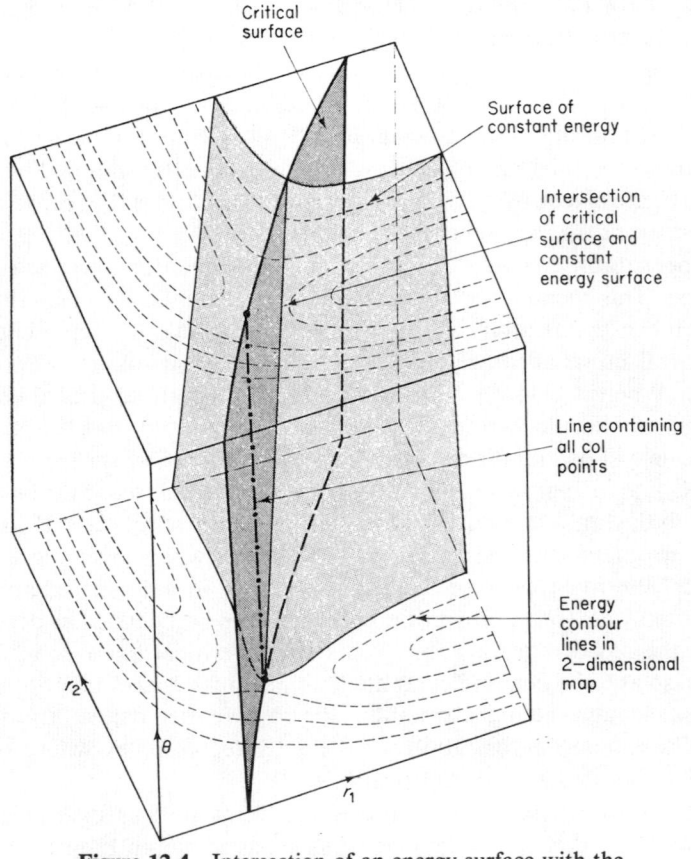

Figure 12.4 Intersection of an energy surface with the critical surface for a reaction involving a triatomic system requiring three internal coordinates r_1, r_2 and θ.

surface, and one of the constant energy surfaces intersecting it at right angles. Successful reaction is now said to be achieved when the representative point moving over the three-dimensional PE surface passes through the two-dimensional critical surface. All configurations represented by points on the critical surface are termed transition states.

It is obvious that it becomes increasingly difficult to visualize the PE diagram as the number of dimensions increases, but it should be clear from the above arguments that it is our imagination not the basic theory which is limited to a small number of dimensions. The theory should therefore be applicable to the reactions of molecules containing many atoms. If n dimensions are required to represent the internal configuration of the complex, then the PE diagram will require $(n + 1)$ dimensions for its representation. The "col" in the diagram will be an $(n - 2)$-dimensional manifold on the n-dimensional PE manifold (or surface). The critical surface will be an $(n - 1)$-dimensional manifold which will contain the manifold of the col, and will be perpendicular everywhere to the constant energy manifolds, also $(n - 1)$-dimensional.

The problem of obtaining a theoretical expression for a reaction rate now resolves itself into evaluating how fast molecules pass across the critical line or surface in the PE diagram. The problem is in two distinct parts. The first is the calculation of the shape of the PE surface, and the second calculation of the rate at which transition state complexes cross the surface. The first problem is one for quantum mechanics and is generally insoluble except in the simplest cases where three simple atoms are involved, say two H atoms and the atom of a first row element. To solve the second problem is simpler. Two assumptions are required in addition to those which are normally made in molecular thermodynamics.

Assumption 1

The velocity component of the motion of the representative point perpendicular to the critical surface is assumed to behave exactly as if it were a one-dimensional translation in free space. Motion in this direction is called "motion in the reaction coordinate": the "reaction coordinate" is thus the perpendicular distance of any point from the critical surface. The term is often used for the distance measured along the minimum energy path, but this definition is very restrictive. Figure 12.5 illustrates the idea for a two-dimensional PE surface. This assumption holds irrespective of whether the representative point passes over the critical surface near the col or far from it. In justification of the assumption we note (a) that because of the way in which the critical surface is defined, any line perpendicular to it must lie within a constant energy contour at the point of crossing: motion in a constant potential field is one of the characteristics of free translation, and (b) that, as shown above for the H—Cl—Cl system, translational motion may indeed be directly represented by motion of the representative point when this motion lies within the appropriate part of the PE diagram.

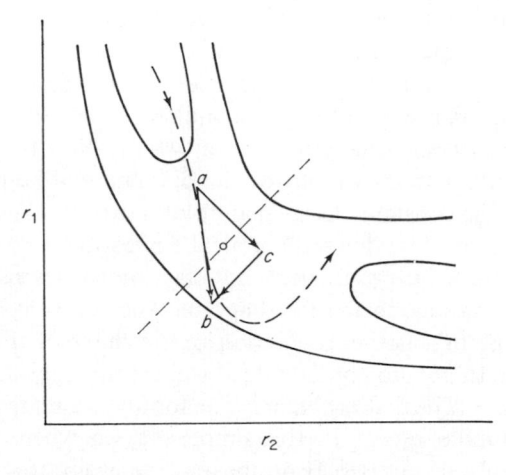

Figure 12.5 Components of motion of the representative point in the region of the critical line. ab = vector representing motion of representative point; ac = vector giving component perpendicular to critical line; cb = vector giving component of motion within critical line.

It is thus not unreasonable to assume that motion along a constant energy contour elsewhere on a PE surface may be considered to obey the same laws, provided that the extent of the region of constant potential energy is sufficient that quantum effects are not serious. If the curvature is great and the barrier very narrow, quantum mechanical tunnelling will occur and the above assumption no longer holds accurately.

The motion in the reaction coordinate may then be separated from motions within the critical surface in the same way as free translational motion may be regarded as independent of the internal motion of molecules. For example motion along oxc in Figure 12.2 requires the simultaneous extension of r_1 and r_2, and therefore represents the symmetrical stretching mode of H—Cl—Cl. It is easy to see that this type of motion produces no contribution to reaction. Motion in the reaction coordinate, that is perpendicular to oxc, requires the contraction of r_1 and simultaneous extension of r_2. For a normal molecule this would represent the unsymmetrical stretching mode, but since for the complex there is no restoring force (the PE surface is flat in the direction of motion) this mode represents reaction.

Assumption 1 enables a simple expression to be developed for the reaction rate. Suppose that the representative points for the systems in a certain group pass over the critical line or surface with velocities, s, in the reaction coordinate. How do we obtain the rate of passage of these systems over the line? The problem is like that of counting cars passing over a bridge, each car being observed only by means of the impulse it produces passing over a control line. Suppose that we are interested in cars which travel at around 30 m.p.h., and that on average it is known (independently) that there are 10 such cars per mile of road leading to the bridge. In the course of 1 hour the cars on 30 miles of road will reach and pass over the bridge. The rate of passage is clearly 30×10 cars per hour. Generalizing for any type of system passing a check point we obtain:

$$\text{(rate of passage)} = \text{(speed, } s) \times \text{(number of systems with speed, } s, \text{ per unit length of path)}$$
$$(12.0.5)$$

Assumption 2

The second assumption allows molecular thermodynamics, developed for assemblies in equilibrium, to be applied to assemblies which are far from equilibrium. According to the principle of detailed balancing, when any chemical reaction is at complete equilibrium, the average number of systems passing forwards through any defined intermediate or transition state (however wide or narrow the definition) is equal within statistical limits to the number passing in the reverse direction. We make the assumption that the systems passing forwards and backwards through any transition state are independent, and that complete or partial removal of the systems coming from the product side in any reaction, will have no influence on the number passing in the forward direction. This means that the forward rate can be found even when the products are completely absent. Enlarging on equation (12.0.5) we can now write

(rate of forward = $(\frac{1}{2})$ × (speed, s) ×
reaction)   (number of systems per unit length of reaction coordinate with speed s, which are present at complete equilibrium between reactants and products)   (12.0.6)

In many treatments a "transmission coefficient", κ, is included in equation (12.0.6) and those derived from it to allow for a partial failure of this assumption and from any effects of quantum mechanical tunnelling. For some special cases it is possible to evaluate κ, but it is essentially a correction factor of a rather arbitrary nature and we shall omit it.

The total rate of reaction is finally obtained by summing the rate of forward passage for all speeds, making proper allowance according to molecular thermodynamics for the distribution of the different speeds.

12.1 DERIVATION OF THE ABSOLUTE RATE EQUATION

We consider the general bimolecular group exchange reaction, which for complete equilibrium, is written

$$A + B \rightleftharpoons X^{\ddagger} \rightleftharpoons C + D \qquad (12.1.1)$$

where $X^{\ddagger}$ is the activated complex. We concentrate initially on those complexes whose speeds in the reaction coordinate lie in the range s to $s + ds$. According to equation (12.0.6) their contribution to the total reaction rate will be

$$R_{(s \text{ to } s + ds)} = \tfrac{1}{2}sc^{*}_{(s \text{ to } s + ds)} \qquad (12.1.2)$$

where $R_{(s \text{ to } s + ds)}$ is the reaction rate (in systems per unit volume per unit time), $c^{*}_{(s \text{ to } s + ds)}$ is the equilibrium concentration of complexes per unit length of reaction coordinate at the critical surface with speeds in the range s to $s + ds$. The concentration c^{*} may be obtained by the methods of molecular thermodynamics, and is given by

$$\frac{c^{*}_{(s \text{ to } s + ds)}}{c_A c_B} = \frac{q^{\circ *}_{(s \text{ to } s + ds)}}{q^{\circ}_A q^{\circ}_B} \exp \left[\frac{-\Delta \epsilon^{\ddagger}_0}{kT} \right] \qquad (12.1.3)$$

where $q^{\circ *}_{(s \text{ to } s + ds)}$ is the partition function per unit volume of gas phase per unit length of reaction coordinate, and $\Delta \epsilon^{\ddagger}_0$ is the difference in ground state energy between the reactants and the complex, that is the energy difference between points b and x in Figure 12.2. By the multiplication theorem, since motion in the reaction coordinate is independent of motions in other orthogonal directions, the partition function for the reaction mode may be factored out of $q^{\circ\ddagger}$ giving:

$$q^{\circ*}_{(s \text{ to } s+ds)} = q^{\circ\ddagger} \times q^{\circ,rc}_{(s \text{ to } s+ds)} \qquad (12.1.4)$$

where $q^{\circ,rc}_{(s \text{ to } s+ds)}$ is the partition function per unit length of the reaction coordinate for those systems with velocities in the range s to $s + ds$, and $q^{\circ\ddagger}$ is the partition function for the remaining modes of motion of the complex, that is those modes which are given by movement of the representative point within the critical line or surface. Since the energy in the reaction coordinate, ϵ_{rc}, is the same for all states having velocities in the range s to $s + ds$, the partition function takes the simple form:

$$q^{\circ,rc}_{(s \text{ to } s+ds)} = \begin{Bmatrix} \text{number of translational} \\ \text{states in speed range} \\ (s \text{ to } s + ds) \end{Bmatrix} \exp\left[-\frac{\epsilon_{rc}}{kT}\right] \qquad (12.1.5)$$

The number quantum states for one-dimensional translation in unit length with a speed range ds is obtained from the quantization rule (see Chapter 2)

$$p = \mu s = hn/2a, \quad a = 1 \qquad (12.1.6)$$

where μ is the reduced mass for motion in the reaction coordinate. The number of states is then

$$dn = (2\mu/h)\, ds \qquad (12.1.7)$$

and the energy in the reaction coordinate is $\epsilon_{rc} = \mu s^2/2$. The partition function for the reaction coordinate is then

$$q^{\circ,rc}_{(s \text{ to } s+ds)} = \frac{2\mu}{h} \exp\left[-\frac{\epsilon_{rc}}{kT}\right] ds \qquad (12.1.8)$$

Inserting (12.1.8) into (12.1.4), and the result into (12.1.3) gives after minor rearrangement:

$$c^{*}_{(s \text{ to } s+ds)} = \frac{q^{\circ\ddagger}}{q^{\circ}_A q^{\circ}_B} c_A c_B \exp\left[-\frac{\Delta\epsilon^{\ddagger}_0}{kT}\right] \times \frac{2\mu}{h} \exp\left[-\frac{\epsilon_{rc}}{kT}\right] ds \qquad (12.1.9)$$

Inserting (12.1.9) into (12.1.2) gives the contribution to the reaction rate:

$$R_{(s \text{ to } s+ds)} = \frac{s}{2} \frac{q^{\circ\ddagger}}{q_A^\circ q_B^\circ} c_A c_B \exp\left[-\frac{\Delta\epsilon_0^\ddagger}{kT}\right] \frac{2\mu}{h} \exp\left[-\frac{\mu s^2}{2kT}\right] ds \quad (12.1.10)$$

Rearranging slightly and integrating gives the total reaction rate:

$$R = \frac{1}{h} \left\{\frac{q^{\circ\ddagger}}{q_A^\circ q_B^\circ}\right\} c_A c_B \exp\left[-\frac{\Delta\epsilon_0^\ddagger}{kT}\right] \int_0^\infty \exp\left[-\frac{\mu s^2}{2kT}\right] \mu s \, ds \quad (12.1.11)$$

Since $\mu s \, ds = d(\mu s^2/2)$, the integral has the form $\int_0^\infty \exp\left[-X/kT\right] dX$ $= kT$ leading finally to the result:

$$R = \frac{kT}{h} \left\{\frac{q^{\circ\ddagger}}{q_A^\circ q_B^\circ}\right\} \exp\left[-\frac{\Delta\epsilon_0^\ddagger}{kT}\right] c_A c_B \quad (12.1.12)$$

The rate constant for the reaction is then

$$k_{bi} = \frac{kT}{h} \frac{q^{\circ\ddagger}}{q_A^\circ q_B^\circ} \exp\left[-\frac{\Delta\epsilon_0^\ddagger}{kT}\right] = \frac{kT}{h} K^\ddagger \quad (12.1.13)$$

In the final form of equation (12.1.13), the partition functions and exponential factor have been replaced by $K^\ddagger$ which may be formally regarded as the equilibrium constant for the formation of the transition state complex. $K^\ddagger$ is a purely theoretical quantity which is to be evaluated by the methods of molecular thermodynamics from the partition functions of the reactants and the complex. In evaluating the partition function for the complex the reaction coordinate is excluded, as this type of motion has already been allowed for in the derivation of the factor kT/h. For any polyatomic complex $q^{\circ\ddagger}$ is evaluated using $3n-6$ internal vibrational modes when the complex is linear, and $3n-7$ internal modes (vibrational and internal rotational) if non-linear. The factor kT/h is a universal frequency factor whose value at 500 K is

$$kT/h = 1\cdot38 \times 10^{-23} \times 5 \times 10^2/(6\cdot65 \times 10^{-34})$$

$$= 1\cdot04 \times 10^{13} \text{ s}^{-1} \quad (12.1.14)$$

12.2 BIMOLECULAR REACTIONS: CORRELATION OF COLLISION AND TRANSITION STATE THEORY

According to simple collision theory, elementary bimolecular reactions occur as a direct result of molecular collisions if the molecules are properly orientated, and if the collisions are sufficiently violent.

The simplest model of the collision process regards the molecules as hard spheres. The fraction of collisions for which the molecules are properly orientated is then a constant which is independent of the violence of the collision. For instance if two spheres have fractions f_A and f_B of their surfaces painted, then the fraction of all collisions in which the two painted parts impinge will be $f_{AB} = f_A f_B$. In collision theory this fraction is called the steric factor and usually denoted by P (not to be confused with P used to denote probability in Chapter 4).

The violence of a collision between spherical particles is related to the amount of energy which can be released at the moment of impact should the spheres for instance fail to behave as hard perfectly elastic spheres. This energy can be shown to be given by:

$$\epsilon_{rel} = \mu v_{rel}^2/2 \qquad (12.2.1)$$

where μ is the reduced mass of the colliding pair, and v_{rel} is their relative velocity of approach along the line of centres at the moment of impact.

It can be shown by arguments which are somewhat similar to, but more complex than, those used in the derivation of the ideal gas pressure, that the rate of collisions between spheres where ϵ_{rel} exceeds a minimum value ϵ_0, is given by equation (12.2.2) (see for example Fowler and Guggenheim, *Statistical Thermodynamics*, p. 491; Benson, *Foundations of Chemical Kinetics*, McGraw Hill, p. 148)

$$Z_{AB}(\epsilon_0) = \pi D_{AB}^2 (8kT/\pi\mu)^{1/2} \exp\left[-\epsilon_0/kT\right] c_A c_B \qquad (12.2.2)$$

where D_{AB} is the mean collision diameter of the colliding pair, that is $D_{AB} = (D_A + D_B)/2$. The factor πD_{AB}^2 is called the collision cross section for the encounter. The standard collision number is defined as the total collision rate irrespective of energy at unit concentrations of A and B. It is obtained from (12.2.2) by setting $c_A = c_B = 1$, and $\epsilon_0 = 0$.

$$Z_{AB}^\circ = \pi D_{AB}^2 (8kT/\pi\mu)^{1/2} \qquad (12.2.3)$$

Typically D_{AB} is about 5×10^{-10} m. If the molecular weights of A and B are about 60, then $\mu = 30/(6 \times 10^{26}) = 5 \times 10^{-26}$ kg. For a temperature of 500 K the standard collision rate is then

$$Z_{AB}^\circ = 3.1 \times 25 \times 10^{-20} \left\{\frac{8 \times 1.4 \times 10^{-23} \times 5 \times 10^2}{3.1 \times 5 \times 10^{-26}}\right\}^{1/2}$$

$$= 4.5 \times 10^{-16} \text{ m}^3 \text{ s}^{-1} \qquad (12.2.4)$$

The standard collision number is often expressed in molar rather than molecular units. If the standard concentration is 1 mole per unit volume and the collision rate is measured in moles of collisions per unit time, the transformation is made by multiplying the standard collision number by 6.02×10^{23}. This gives:

$$Z_{AB}^{\circ} = 2.7 \times 10^8 \; mol^{-1} \, m^3 \, s^{-1}$$

$$= 2.7 \times 10^{11} \; mol^{-1} \, l \, s^{-1}$$

$$= 2.7 \times 10^{14} \; mol^{-1} \, cm^3 \, s^{-1} \qquad (12.2.5)$$

Inserting the steric factor into equation (12.2.2) and using (12.2.3) for the standard collision number gives the reaction rate according to simple collision theory as:

$$Rate = PZ_{AB}^{\circ} \exp \left[-\epsilon_0/kT \right] c_A c_B \qquad (12.2.6)$$

and the rate constant as:

$$k_{bi} = PZ_{AB}^{\circ} \exp \left[-\epsilon_0/kT \right] \qquad (12.2.7)$$

Equation (12.2.7) differs so profoundly from equation (12.1.13), which was derived from transition state theory, that it is desirable to examine whether the two theories can be correlated when applied to the one type of reaction for which both can give explicit expressions for k_{bi}. This is the reaction between structureless spherical particles for which there is no orientational requirement, that is $P = 1$. For such a reaction collision theory gives rise to equation (12.2.2).

The appropriate expression from transition state theory is found by evaluating the partition functions in equation (12.1.13). When the reacting molecules are structureless spheres, internal modes of motion must be unimportant in the formation of the complex, and the molecules must be regarded as simple atoms. The partition functions for the reactants, A and B, are thus:

$$q_A^{\circ} = (2\pi m_A kT/h^2)^{3/2}$$

$$q_B^{\circ} = (2\pi m_B kT/h^2)^{3/2} \qquad (12.2.8)$$

The transition state complex is diatomic, and if a normal molecule would possess two degrees of rotational freedom, and one of vibrational freedom. The vibrational mode is however the reaction mode, and so is omitted in evaluating $q^{\circ\ddagger}$. Thus we obtain

$$q^{\circ\ddagger} = \left\{\frac{2\pi(m_A + m_B)kT}{h^2}\right\}^{3/2} \times \frac{8\pi^2\mu R^2 kT}{h^2} \qquad (12.2.9)$$

where R is the internuclear distance in the complex, and μR^2 is the moment of inertia of the complex (see equation 6.2.15). The rate constant is thus given by equation (12.2.10).

$$k_{bi} = \frac{kT}{h}\left\{\frac{2\pi(m_A + m_B)kT/h^2}{(2\pi m_A kT/h^2)(2\pi m_B kT/h^2)}\right\}^{3/2} \times \frac{8\pi^2\mu R^2 kT}{h^2} \exp\left[-\frac{\Delta\epsilon_0^{\ddagger}}{kT}\right]$$

$$= \pi R^2 \left\{\frac{8kT}{\pi\mu}\right\}^{1/2} \exp\left[-\frac{\Delta\epsilon_0^{\ddagger}}{kT}\right] \qquad (12.2.10)$$

The final expression is seen to become identical to (12.2.2) if $R = D_{AB}$, and $\Delta\epsilon_0^{\ddagger} = \epsilon_0$. Both identifications are entirely reasonable, and indeed could hardly be avoided. The two theories, apparently differing widely at first sight give the same results when applied to the simplest possible model. In a sense therefore the transition state theory may be regarded as a sophistication of the collision theory. The two theories differ chiefly in that the simple collision theory contains the arbitrary parameter P, whereas transition state theory, at least in principle, gives a complete expression for the rate constant in terms of molecular parameters. By comparison of the rate constant calculated by transition state theory with equation (12.2.10), the effective steric factor may be found.

Having established the agreement between the two rate theories for the simplest case we examine qualitatively how the value for the steric factor may be predicted by transition state theory for more complex reactions. We first note the orders of magnitude of certain key quantities:

Translational partition function for motion in 1 dimension
$$q_t \approx 10^{10.7}\ \mathrm{m}^{-1}$$

Rotational partition function for motion in one dimension
$$q_r \approx 10^{1.0}$$

Vibrational partition function for a single mode
$$q_v \approx 10^{0.25}$$

kT/h at 500 K $\qquad\qquad kT/h = 10^{13.0}\ \mathrm{s}^{-1}$

$$(12.2.11)$$

These figures are accurate to within factors of about 3 for most cases of interest. Thus in making order of magnitude calculations we need take little account of individual molecular parameters: all translational partition functions, for instance, may be taken to be the same. Applying this idea to equation (12.2.10) we may write

$$k_{bi} = \frac{kT}{h} \left\{ \frac{q_t^3 q_r^2}{q_t^3 q_t^3} \right\} \exp \left[-\frac{\Delta\epsilon_0^{\ddagger}}{kT} \right] = \frac{kT}{h} \times (q_r^2 q_t^{-3}) \exp \left[-\frac{\Delta\epsilon_0^{\ddagger}}{kT} \right]$$

$$= (10^{13} \times 10^2 \times 10^{-32}) \exp \left[-\frac{\Delta\epsilon_0^{\ddagger}}{kT} \right] = 10^{-17} \exp \left[-\frac{\Delta\epsilon_0^{\ddagger}}{kT} \right]$$

$$\tag{12.2.12}$$

The pre-exponential factor 10^{-17} compares reasonably well with that given by exact calculation in (12.2.4).

We now consider the reaction of an atom with a diatomic molecule to form a linear transition state, for example reaction (12.0.1). The general reaction is

$$A + BC \rightarrow A\text{—}B\text{—}C^{\ddagger} \rightarrow AB + C \tag{12.2.13}$$

The order of magnitude partition functions to be inserted into (12.1.13) are then

$$\left.\begin{array}{lll} \text{Atom, A} & q_A^{\circ} & = q_t^3 \\ \text{Diatom AB} & q_{AB}^{\circ} & = q_t^3 q_r^2 q_v \\ \text{Linear triatomic complex } ABC^{\ddagger} & q^{\circ\ddagger} & = q_t^3 q_r^2 q_v^3 \end{array}\right\} \tag{12.2.14}$$

In dealing with the complex only three of the normal four vibrational modes are included since the unsymmetrical stretching mode is taken as the reaction mode. The rate thus becomes

$$k_{bi} = \frac{kT}{h} \left\{ \frac{q_t^3 q_r^2 q_v^3}{q_t^3 q_t^3 q_r^2 q_v} \right\} \exp \left[-\frac{\Delta\epsilon_0^{\ddagger}}{kT} \right]$$

$$= \left\{ \frac{kT}{h} (q_t^{-3} q_r^2) \exp \left[-\frac{\Delta\epsilon_0^{\ddagger}}{kT} \right] \right\} \left(\frac{q_v}{q_r} \right)^2$$

$$= \left\{ Z^{\circ} \exp \left[-\frac{\Delta\epsilon_0^{\ddagger}}{kT} \right] \right\} \left(\frac{q_v}{q_r} \right)^2 \tag{12.2.15}$$

The first factor in equation (12.2.15) is the right-hand side of equation

(12.2.12) line 2. The second factor may therefore be identified with the steric factor of simple collision theory. Thus

$$P = (q_v/q_r)^2 \approx 10^{-1.5} \qquad (12.2.16)$$

Table 12.1 summarizes the results of similar analysis applied to other types of bimolecular reaction. Examination of column 8 shows that the steric factor always has the form

$$P = (q_v/q_r)^n \qquad (12.2.17)$$

where n is an integer between 0 and 5. This result implies that in molecular thermodynamic terms the steric factor arises through the replacement of rotational modes of motion by vibrational modes when the reactants combine to form the transition state complex. P is below unity because of the relative sparseness of vibrational compared to rotational quantum states.

The numerical values of the steric factors given in Table 12.1 are subject to large error, and must be taken only as a rough guide. The principle which they establish is that the steric factor becomes lower as the reactants become more complex. This is just what would be expected on the basis of an orientational requirement. Quite often the new vibrations which arise in the complex as a result of loss of rotational modes in the reactants have rather low frequencies since they are associated with motions about partially formed bonds. The ratio (q_v/q_r) for some or all of these modes may then be considerably larger than 0·2. Another factor which often raises the steric factor is that certain modes of vibration which are relatively stiff in the reactants become "softened" in forming the complex because of the weaker bonding in the complex. Thus one q_v in the reactants is replaced by a larger q_v in the complex.

On the other hand many large molecules have rotational partition functions which are greatly in excess of 10 per degree of freedom. With such systems smaller steric factors are expected. Finally when nearly free internal rotational modes are converted into vibrational modes in the complex, say by the formation of a ring in the transition state, lowering of the steric factor is expected. About the most that can be claimed regarding the values of P given in Table 12.1 is that $\log_{10} P$ for a given class of reactions is likely to be within a factor of two of an experimental value.

Table 12.1 Semi-quantitative steric factors for bimolecular reactions

Reactant molecules			Complex	Partition functions				Steric factor, P[a]	
A	+	B	$\rightarrow$ AB‡	q_A°	q_B°	$q^{\circ‡}$	$q^{\circ‡}/q_A^\circ q_B^\circ$	Ratio, P	Value, P
Atom	+ Atom		$\rightarrow$ Diatom	q_t^3	q_t^3	$q_t^3 q_r^2$	$q_t^{-3} q_r^2$	1	1
Atom	+ Diatom		$\rightarrow$ nl-Triatom[b]	q_t^3	$q_t^3 q_r^2 q_v$	$q_t^3 q_r^3 q_v^2$	$q_t^{-3} q_r q_v$	(q_v/q_t)	$10^{-0.7}$
Atom	+ Diatom		$\rightarrow$ l-Triatom	q_t^3	$q_t^3 q_r^2 q_v$	$q_t^3 q_r^2 q_v^3$	$q_t^{-3} q_v^2$	$(q_v/q_t)^2$	$10^{-1.5}$
Diatom	+ Diatom		$\rightarrow$ nl-Tetra-atom	$q_t^3 q_r^2 q_v$	$q_t^3 q_r^2 q_v$	$q_t^3 q_r^3 q_v^5$	$q_t^{-3} q_r^{-1} q_v^3$	$(q_v/q_r)^3$	$10^{-2.2}$
Atom	+ nl-Polyatom		$\rightarrow$ nl-Polyatom n atom	q_t^3	$q_t^3 q_r^3 q_v^{3n-6}$	$q_t^3 q_r^3 q_v^{3n-4}$	$q_t^{-3} q_v^2$	$(q_v/q_t)^2$	$10^{-1.5}$
Atom	+ nl-Polyatom n atoms		$\rightarrow$ nl-Polyatom $n + 1$ atoms	q_t^3	$q_t^3 q_r^3 q_v^{3n-6}$	$q_t^3 q_r^3 q_v^{3n-3}$	$q_t^{-3} q_v^3$	$(q_v/q_t)^2$	$10^{-1.5}$
Diatom	+ nl-Polyatom n atoms		$\rightarrow$ nl-Polyatom $n + 2$ atoms	$q_t^3 q_r^2 q_v$	$q_t^3 q_r^3 q_v^{3n-6}$	$q_t^3 q_r^3 q_v^{3n-1}$	$q_t^{-3} q_r^{-2} q_v^4$	$(q_v/q_r)^4$	$10^{-3.0}$
nl-Polyatom n atoms	+ nl-Polyatom m atoms		$\rightarrow$ nl-Polyatom $n + m$ atoms	$q_t^3 q_r^3 q_v^{3n-6}$	$q_t^3 q_r^3 q_v^{3m-6}$	$q_t^3 q_r^3 q_v^{3(n+m)-7}$	$q_t^{-3} q_r^{-3} q_v^5$	$(q_v/q_r)^5$	$10^{-3.7}$

[a] $P = (kT/h)\{q^{\circ‡}/(q_A^\circ q_B^\circ)\}(1/Z^\circ)$ = theoretical value for steric factor.

[b] nl $\equiv$ non-linear molecule, l $\equiv$ linear molecule or complex.

Table 12.2 gives an indication of the range of steric factors which are found experimentally. They cover within an order of magnitude or two those predicted by the semi-quantitative approach. The steric factors for reactions of atoms with molecules are normally between 0·1 and

Table 12.2 Arrhenius parameters and steric factors for bimolecular reactions

Reaction	$\log_{10} \dfrac{A}{m^3\ mol^{-1}s^{-1}}$	$\dfrac{E}{kJ\ mol^{-1}}$	P
Atoms			
$Cl + H_2 \rightarrow HCl + H$	7·9	22·8	0·2
$Cl + C_2H_6 \rightarrow HCl + C_2H_5$	7·8	4·3	0·2
$Cl + C_2H_2Cl_2 \rightarrow C_2H_2Cl_3$	7·4	0·0	0·1
$Cl + CHCl_3 \rightarrow HCl + CCl_3$	6·8	13·8	0·03
$H + C_2H_4 \rightarrow C_2H_5$	7·4	14·0	0·03
Diatomics			
$OH + CO \rightarrow CO_2 + H$	5·6	4·5	$1·5 \times 10^{-3}$
$OH + H_2 \rightarrow H_2O + H$	7·3	21·5	0·04
$OH + CH_4 \rightarrow H_2O + CH_3$	7·5	23·0	0·1
$NO + O_3 \rightarrow NO_2 + O_2$	6·0	10·0	3×10^{-3}
$HI + C_2H_4 \rightarrow C_2H_5I(ring)^a$	5·5	120	9×10^{-4}
Polyatomics			
$CH_3 + CH_4 \rightarrow CH_4 + CH_3$	5·2	53	4×10^{-4}
$CH_3 + C_2H_6 \rightarrow CH_4 + C_2H_5$	5·5	45	7×10^{-4}
$2C_2F_4 \rightarrow$ cyclo $C_4F_8(ring)$	4·6	104	1×10^{-4}
2(1,3-pentadiene) $\rightarrow$ dimer (ring)	4·3	106	2×10^{-5}
2(cyclopentadiene) $\rightarrow$ dimer (ring)	1·7	61	8×10^{-8}

^{a} (Ring) means that the transition state is a ring. The last two reactions are Diels–Alder additions of the form unity. Those for reactions of diatoms with polyatoms are somewhat smaller and cover the range 10^{-3} to $0\cdot1$, Steric factors for atom transfer reactions between polyatoms (for example $CH_3 + RH = CH_4 + R$) are generally in the region of 10^{-3}. Very low steric factors occur when extensive loss of internal rotation or of low frequency vibrations occurs due to the formation of a ring transition state.

Precise calculation of pre-exponential factors for bimolecular reactions requires detailed knowledge of the molecular parameters of reactants and complexes, for only then can the partition functions in equation (12.1.13) be evaluated accurately. Generally the molecular parameters for reactants are reasonably well established, or can at least be estimated with some confidence by analogy with molecules of established structure. The molecular parameters for transition state complexes can never be directly determined. They can only be estimated. This usually presents problems since the frequencies for those modes of vibration which have sizable amplitudes near the seat of reaction can be only roughly estimated: their frequencies are inevitably lower than those of similar modes in normal molecules, but how much lower is uncertain. Calculations of high accuracy are thus not possible. Nevertheless the reasonable correlations which have been obtained between calculated and experimental values show that useful information about the nature of transition state complexes can be obtained by this kind of comparison. An example of this type of calculation is found in a paper by Herschbach, Johnston, Pitzer and Powell, *J. Chem. Physics*, **25**, 736 (1956).

The temperature dependence of experimentally determined rate constants is generally expressed by the Arrhenius equation

$$k_{expt} = A_{expt} \exp\left[-E_{expt}/RT\right] \qquad (12.2.18)$$

A_{expt} is called the experimental frequency factor or more simply the "A-factor"; E_{expt} is the experimental activation energy. The two parameters are obtained by plotting $\log_{10} k_{expt}$ against $1/T$. The gradient of the line is then $-E_{expt}/2\cdot303R$, and the intercept on the $\log_{10} k_{expt}$ axis is $\log_{10} A_{expt}$. The experimental activation energy can then be defined by the equation

$$E_{expt} = RT^2 \, d(\ln k_{expt})/dT \qquad (12.2.19)$$

Theoretical expressions for rate constants have the general form:

$$k_{th} = B(T) \exp\left[-W/RT\right] \qquad (12.2.20)$$

where $B(T)$ is a relatively weak function of temperature compared to the exponential factor, W being some energy term. For example, according to simple collision theory $B(T)$ is proportional to $T^{1/2}$ and W is the minimum energy for reaction. According to transition state theory $B(T)$ will generally be proportional to T raised to some small positive power, and W will be the difference in ground state energies between the reactants and the complex.

The theoretical quantity which has to be compared with E_{expt} is obtained by applying equation (12.2.19) to (12.2.20):

$$E_{th} = RT^2 \, d \, (\ln B(T))/ \, dT + W$$
$$= \theta RT + W \qquad (12.2.21)$$

where $\theta = T \, d \, (\ln B(T))/dT$. If $B(T) = CT^n$, then $\theta = n$.

Rewriting equation (12.2.20) we then obtain

$$k_{th} = B(T) \exp [\theta] \exp [-E_{th}/RT] \qquad (12.2.22)$$

The theoretical quantity which has to be compared with A_{expt} is then

$$A_{th} = B(T) \exp [\theta] \qquad (12.2.23)$$

Since θ is generally small the correction factor is also small, but it should not be ignored.

Table 12.3 compares the experimental and theoretical A-factors for a number of bimolecular reactions. The theoretical values are those calculated by Herschbach et al. The experimental values have been updated in the light of work since 1956, and are taken from a compilation by Kondratiev, *Velocity Constants of Gas Reactions*, Academy of Sciences, U.S.S.R., Moscow, 1970. For most of the reactions the transition state theory gives reasonable agreement with experiment. Where there are discrepancies they are mostly in the direction that the calculated values are too low. This is to be expected since vibration frequencies are likely to be lower in the complex than in normal molecules of similar structure. In making comparisons between experiment and theory it must be remembered that the experimental A-factors may themselves be considerably in error, for they are generally obtained by quite lengthy extrapolation of plots of $\log_{10} k_{expt}$ against $1/T$. Even when the experimental data are themselves reliable and accurate, the Arrhenius parameters may have large errors, especially if the temperature range over which k is measured is small. Typically the errors in

Table 12.3 Comparison of experimental A-factors with values calculated by transition state theory[a]

Reaction	Temperature range/K	E_{expt} kJ mol^{-1}	$10^{-6} A_{expt}$ m^3 mol^{-1} s^{-1}	$10^{-6} A_{calc}$ m^3 mol^{-1} s^{-1}	$10^{-6} Z°$ m^3 mol^{-1} s^{-1}	$P = A_{expt}/Z°$	A_{calc}/A_{expt}
NO + O$_3$ → NO$_2$ + O$_2$	216–322	17·5	0·85	0·44	47	0·018	0·5
NO$_2$ + O$_3$ → NO$_3$ + O$_2$	286–302	29·3	5·9	0·14	63	0·095	0·025
NO$_2$ + F$_2$ → NO$_2$F + F	300–343	43·9	1·6	0·12	59	0·027	0·07
NO$_2$ + CO → NO + CO$_2$	500–800	123	1·9	6·0	74	0·026	3
2NO$_2$ → 2NO + O$_2$	473–1020	113	4·1	4·5	43	0·105	1·1
NO + NO$_2$Cl → NOCl + NO$_2$	274–344	28·8	0·83	0·84	71	0·012	1·0
2NOCl → 2NO + Cl$_2$	373–1020	98·5	5·3	0·44	59	0·090	0·08
NOCl + Cl → NO + Cl$_2$	298–328	4·6	11·4	4·4	57	0·20	0·4
NO + Cl$_2$ → NOCl + Cl	430–673	83	2·8	1·2	93	0·03	0·4
F$_2$ + ClO$_2$ → FClO$_2$ + F	227–247	35·4	0·04	0·08	47	0·001	2
2ClO → Cl$_2$ + O$_2$	300–433	0	0·04	0·01	26	0·0015	0·25

[a] From Herschbach et al., *J. Chem. Physics*, **25**, 736 (1956).

the A-factors for the reactions listed in Table 12.3 will be around a factor of three, except where the T_{max}/T_{min} ratio is large.

12.3 BIMOLECULAR ASSOCIATIONS, UNIMOLECULAR DISSOCIATIONS AND UNIMOLECULAR REARRANGEMENTS

Potential energy diagrams for bimolecular associations differ from those for bimolecular exchange reactions in having no escape valleys on the product sides. The PE diagram for the combination of an atom with a diatom, for example $H + O_2 = HO_2$, where the angle θ is fixed (Figure 12.1) may qualitatively be represented as in Figure 12.6. The

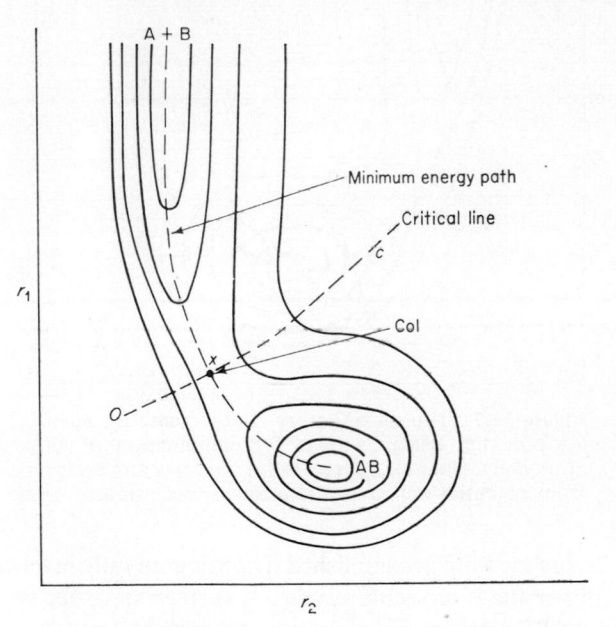

Figure 12.6 Potential energy diagram for combination of an atom with a diatomic molecule requiring only two internal coordinates.

reaction path of minimum energy now terminates in a deep well or "bottle" rather than in a flat east–west valley. When reaction occurs, the representative point passes over the critical line into the region of the well. Since the triatomic system cannot lose its internal energy by converting it into translational energy of two product molecules, the representative point must wander around the well like a bee in a bottle until it eventually finds the only possible escape route over the col and back to the reactants. An example of such a trajectory is shown in Figure 12.7. A bimolecular association such as $A + B \rightarrow AB$ cannot

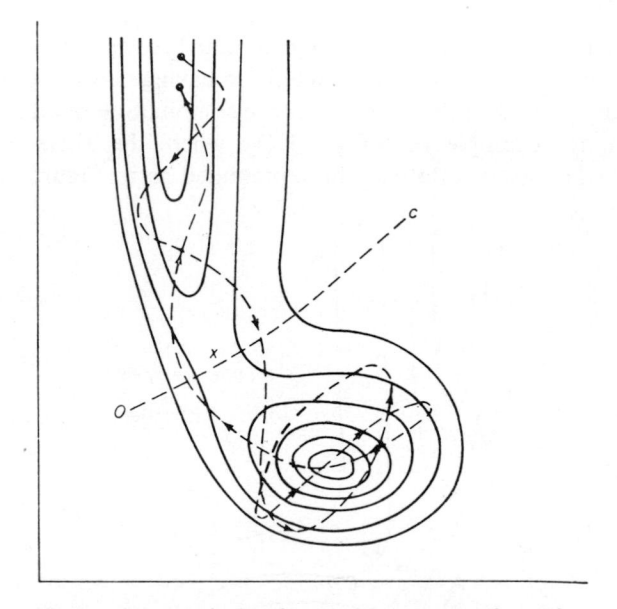

Figure 12.7 Typical trajectory of representative point on potential energy surface for combination of an atom and a diatomic molecule showing entry and escape from potential well after a number of oscillations.

therefore be successfully accomplished if considered only in terms of the PE diagram for the bimolecule system. A further step, not considered in drawing the PE diagram, must intervene to prevent a system which has passed over the critical line or surface from returning.

The transition state theory developed above gives the rate at which reactant systems pass over the critical line or surface in the forward direction, but takes no account, unless by inclusion of an arbitrary transmission coefficient, of any systems which having made the crossing return again to the reactant side. When the representative point has passed into the region of the well there is a finite delay before it normally escapes. This delay is greater the more complex the motions of the molecule and of its representative point. During this period, if the gas pressure is sufficiently high, the excited product molecule may collide with some other unexcited molecule. There is then a high probability (usually greater than 0·1) that sufficient energy will be transferred from the excited molecule that it can no longer escape over the col. It thus becomes trapped in the bottle, and eventually after numerous further collisions becomes a normal thermally equilibrated molecule of product. In practice activated molecules containing six to eight atoms (for example $C_2H_6^*$ formed by association of CH_3 radicals) are efficiently deactivated at pressures of one atmosphere.

Equation (12.1.13) for bimolecular group exchange reactions thus applies to bimolecular association reactions in the so-called "high-pressure region". When the pressure is low the rate constant calculated by elementary transition state theory will be too high.

By the principle of detailed balancing the theory for bimolecular association reactions must, with the same reservations, apply to the reverse process, namely unimolecular dissociation into two product molecules. Consider the general reaction

$$A + B \underset{k_{uni}}{\overset{k_{bi}}{\rightleftharpoons}} X^\ddagger \rightleftharpoons AB \qquad (12.3.1)$$

where k_{bi} and k_{uni} are the high pressure rate constants for the association and dissociation reactions respectively. For equilibrium we can write

$$K_c \equiv \frac{c_{AB}}{c_A c_B} = \frac{k_{bi}}{k_{uni}} \qquad (12.3.2)$$

According to equilibrium theory (Chapter 11) we have

$$K_c = \frac{q_{AB}^\circ}{q_A^\circ q_B^\circ} \exp\left[-\frac{\Delta\epsilon_0^\circ}{kT}\right] \qquad (12.3.3)$$

where $\Delta\epsilon_0^\circ$ is the difference in ground state energies between the reactants and products. By transition state theory we have

$$k_{bi} = \frac{kT}{h} \frac{q^{\circ\ddagger}}{q_A^\circ q_B^\circ} \exp\left[-\frac{\Delta\epsilon_{0,f}^\ddagger}{kT}\right] \qquad (12.3.4)$$

where $\Delta\epsilon_{0,f}^\ddagger$ is the difference in ground state energies between the reactants and the complex. Combination of the last three equations gives after minor rearrangement

$$k_{uni} = \frac{k_{bi}}{K_c}$$

$$= \frac{kT}{h} \frac{q^{o\ddagger}}{q^{o}_{AB}} \exp\left[-\frac{(\Delta\epsilon^{\ddagger}_{0,f} - \Delta\epsilon^{o}_{0})}{kT}\right]$$

$$= \frac{kT}{h} \frac{q^{o\ddagger}}{q^{o}_{AB}} \exp\left[-\frac{\Delta\epsilon^{\ddagger}_{0,b}}{kT}\right] \qquad (12.3.5)$$

where $\Delta\epsilon^{\ddagger}_{0,b}$ is the difference in ground state energies between the product and the complex as shown in Figure 12.8.

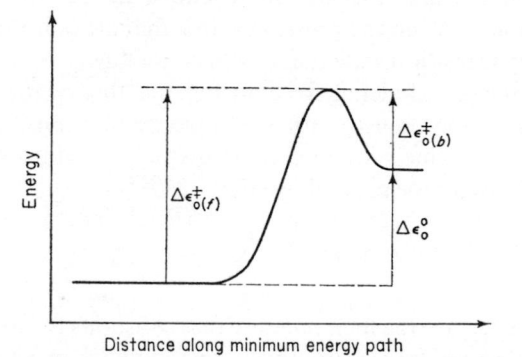

Figure 12.8 Definitions of energies in equations (12.3.3) to (12.3.5).

Equations (12.3.4) and (12.3.5) may be written in the general form which applies in the high pressure region

$$k_r = (kT/h)K^{\ddagger} \qquad (12.3.6)$$

where k_r is the rate constant, and $K^{\ddagger}$ the equilibrium constant calculated by the methods of molecular thermodynamics for the equilibrium between reactants and the complex, the partition function for the complex being calculated assuming that the reaction mode is absent.

It is a simple matter to make the final extension of the treatment to unimolecular rearrangement reactions where the PE diagram consists of two well- or bottle-shaped regions connected by a col or neck. The appropriate diagram for a triatomic system whose internal configuration may be described by two coordinates x and y is shown in Figure 12.9. Without the intervention of some foreign process, any system whose internal energy is sufficient to enable it to pass over the col, will oscillate indefinitely between the two modifications represented by A and B. Such a system will become stabilized as A or B only if it is deactivated, for example, by collision with another molecule or by emission of a

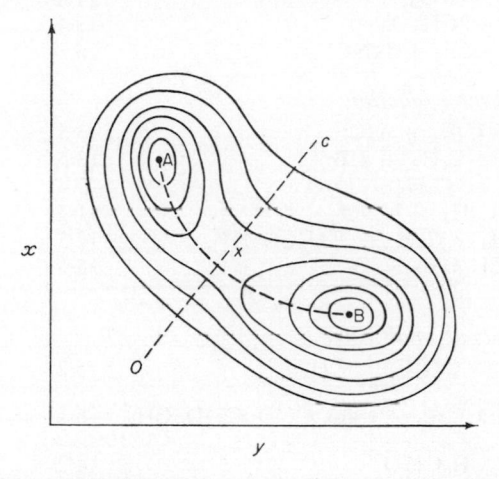

Figure 12.9 Potential energy diagram for a hypothetical unimolecular rearrangement of a triatomic system requiring specification of only two internal coordinates.

light quantum. If the pressure is sufficiently high, molecules with high energy will nearly always be deactivated before they can make a passage over the critical line, or if they have just succeeded in making a passage, before they can return. The transition state theory given above will therefore correctly give the rate of such a reaction in either the forward or reverse directions. For either case equation (12.3.6) will apply.

For unimolecular reactions at the high pressure limit, the pre-exponential or A-factors as shown by the data given in Table 12.4 is of the order of 10^{13} to $10^{15.5}$ s^{-1}. Since kT/h is of the order of 10^{13} s^{-1}, the theory is in qualitative agreement with experiment provided that some loosening of the bonding normally occurs in passing from reactants to complex. On the whole this seems reasonable. We may predict further that relatively low values of unimolecular A-factors will arise when linear molecules form cyclic transition states.

Table 12.4 Arrhenius parameters for some unimolecular reactions [a]

Reaction	$\log_{10}(A/\text{s}^{-1})$	$E/(\text{kJ mol}^{-1})$
Dissociations of single bonds		
$C_2H_6 \to 2CH_3$	16·0	359
$N_2O_5 \to NO_3 + NO_2$	14·8	88
$CH_3OOCH_3 \to 2CH_3O$	15·4	151
$CF_3NNCF_3 \to CF_3 + NNCF_3$	16·2	232
Four and six centre reactions		
$CH_3CH_2Cl \to C_2H_4 + HCl$	13·5	237
$t\text{-}C_4H_9Br \to \text{iso-}C_4H_8 + HBr$	13·5	174
$t\text{-}C_4H_9OH \to \text{iso-}C_4H_8 + H_2O$	13·4	258
$C_2H_5NO_2 \to C_2H_4 + HNO_2$	11·5	174
$CH_3COOC_2H_5 \to C_2H_4 + CH_3COOH$	12·5	200
$CH_2OCH_2OCH_2O \to 3CH_2O$	15·0	198
Rearrangements of small rings		
$CH_2CH_2CH_2 \to CH_3CH{=}CH_2$	15·2	272
cis-$CHD.CHD.CH_2 \to$ trans $CHD.CHD.CH_2$	16·1	273
$CH_2CH_2O \to CH_3CHO$	14·2	238
$CH_2CH{=}CHCH_2 \to CH_2{=}CH.CH{=}CH_2$	13·1	136

[a] Taken from Benson, *Thermochemical Kinetics*, Wiley, 1968.

When the assumption of high pressure of deactivating gas cannot be made the elementary treatment given in Section 12.1 cannot be applied and allowance must be made for the return of systems over the critical line or surface. Since the delay between the initial passage over the col into the bottle, and the subsequent emergence from the bottle depends upon the energy which a molecule contains, while the collision rate to a first approximation does not, the fraction of molecules deactivated will depend upon their energy. This complication has to be allowed for at an early stage in the development of the theory (around equation 12.1.11) The full treatment of the transition state theory of unimolecular reactions and of bimolecular associations is beyond our scope but is dealt with in textbooks on kinetics, for example *Unimolecular Reactions* by Holbrooke and Robinson, Wiley, 1971.

PROBLEMS

12.1 The Lindemann mechanism for the decomposition of cyclobutane may be written, where A = cyclobutane, A* = excited cyclobutane, M = any molecule, E = ethylene.

$$A + M \to A^* + M \qquad \text{(i)}$$

$$A^* + M \to A + M \qquad \text{(ii)}$$

$$A^* \to 2E \qquad \text{(iii)}$$

Classify reactions (i) to (iii) according to the types of basic reaction defined on page 217.

12. Sketch the PE diagrams for exothermic and endothermic bimolecular exchange reactions involving three atoms when θ = constant.

12.3 Sketch the PE diagram for a bimolecular (triatomic) combination reaction $A + BC \to ABC$ involving virtually no activation energy in the forward direction and assuming θ = constant. Draw sections which represent the PE curves for the symmetrical and unsymmetrical stretching mode of ABC. Why cannot a curve be constructed for the bending mode from the PE diagram which has been drawn? Draw similar sections for the symmetrical and unsymmetrical modes in the complex. Which mode is the reaction mode? How does the PE curve for this mode differ from the normal PE curve for a harmonic oscillator?

12.4 Estimate the steric factors for the following reactions, first in terms of $(q_v/q_r)^n$ and then numerically. Thence estimate the pre-exponential factors giving their units.

$$H\cdot + C_2H_4 \to C_2H_5\cdot \qquad \text{(i)}$$

$$CH_3\cdot + CH_3COCH_3 \to CH_4 + CH_3COCH_2\cdot \qquad \text{(ii)}$$

$$OH\cdot + H_2 \to H_2O + H\cdot \qquad \text{(iii)}$$

$$1,3\text{-diene} + \text{olefin} \to \text{cyclohexene} \qquad \text{(iv)}$$

In treating (iv) assume that there is free internal rotation about the single bond in the diene.

12.5 Estimate the pre-exponential factor at 600 K for the following unimolecular reaction which may be taken to involve the loss of two free internal rotational modes of motion in forming the complex:

$$CH_3\cdot\overset{\bullet}{C}H\cdot CH_2\cdot CH_2\cdot CH_2OOH \longrightarrow \begin{array}{c} H_2C-CH_2 \\ | \quad\quad | \\ CH_3\cdot HC \quad CH_2 \\ \diagdown_{}\diagup \\ O \end{array} + OH\cdot$$

12.6 Show that the Arrhenius or experimental activation energy (per mole) for a reaction obeying the rate equation (12.2.2) is $(E_0 + \frac{1}{2}RT)$, and that for the reaction

$$OH\cdot + CH_4 \rightarrow CH_3\cdot + H_2O$$

it would be $(E_0 - \frac{3}{2}RT)$ if the vibrational degrees of freedom in the reactants and complex were unexcited. E_0 in both cases being the minimum energy for reaction at 0 K.

12.7 Pitzer (*J. Amer. Chem. Soc.*, **79**, 1804 (1957)) has estimated that the complex Cl--H--H formed in the reaction $Cl\cdot + H_2 = HCl + H\cdot$ has bonds which are 0·18 Å longer than the normal single bonds in H_2 and HCl (for values see Table 6.2), and that the vibration frequencies in the complex are 1460 cm^{-1} (symmetrical stretch) and 560 cm^{-1} (bend, doubly degenerate). The fundamental vibration frequency in H_2 is 4405 cm^{-1}. The electronic partition function of Cl is obtained from the data given in Problem 11.2. The electronic degeneracy of the complex is 2.

Calculate the pre-exponential factor for the reaction at 500 K, and compare with the experimental value of 8×10^7 mol^{-1} m^3 s^{-1}.

CHAPTER 13

BIBLIOGRAPHY

The following list which is by no means comprehensive gives the titles of a number of texts and reference books which the reader may like to refer to for further information and explanation.

Those called Introductory Texts are written at about the same level as the present book, Nash and Guggenheim being somewhat more elementary, Gurney and Andrews being somewhat more advanced. Those listed as Monographs are considerably more advanced. Tolman is the classic exposition of the foundations of the subject, and Fowler and Guggenheim is probably the most helpful advanced book for chemists. It contains a wealth of illustrative data.

The general textbooks, of course, contain sections on molecular thermodynamics, but are particularly listed since they give a large amount of illustrative data (Partington) or contain tables of functions (Taylor and Glasstone). Herzberg's book is a classic on spectroscopy and is an excellent source book for fundamental vibration frequencies. The books on Reaction Kinetics give accounts of transition state theory, Glasstone, Laidler and Eyring being the classic in this area. Holbrooke and Robinson gives a good account of the transition state theory as applied to basic reactions, and elementary reactions whose rate constants are pressure dependent.

The titles listed under Sources of Data are of two types. Benson and Janz are texts whose object is to show how thermodynamic properties may be calculated. Janz in particular deals with the molecular thermodynamic methods and treats the complications which arise from anharmonicity, restricted rotation, etc. The remaining titles are those of reference books containing tables of thermodynamic data, kinetic data, and molecular data.

Introductory texts

Guggenheim, E. A., *Boltzmann's Distribution Law*, North Holland Publishing Co., Amsterdam, 1963.

Kauzmann, W., *Kinetic Theory of Gases*, Benjamin, 1966.
Nash, L. K., *Elements of Statistical Thermodynamics*, Addison Wesley Publ. Co., 1968.
Rushbrooke, G. S., *Introduction to Statistical Mechanics*, Oxford University Press, 1949.
Gurney, R. W., *Introduction to Statistical Mechanics*, McGraw Hill Book Co., 1949.
Andrews, F., *Equilibrium Statistical Mechanics*, Wiley, 1963.

Monographs

Tolman, R. C., *Principles of Statistical Mechanics*, Oxford University Press, 1938.
Fowler, R. H., and Guggenheim, E. A., *Statistical Thermodynamics*, Cambridge University Press, 1952.
Mayer, J. E., and Mayer, M. G., *Statistical Mechanics*, Wiley, 1940.
Hill, T. L., *Statistical Mechanics*, McGraw Hill Book Co., 1956.
Davidson, N., *Statistical Mechanics*, McGraw Hill Book Co., 1962.
Eyring, H., Henderson, D., Stover, B. J., and Eyring, E., *Statistical Mechanics and Dynamics*, Wiley, 1964.
Wilson, A. H., *Thermodynamics and Statistical Mechanics*, Cambridge University Press, 1966.
Rice, O. K., *Statistical Mechanics, Thermodynamics and Kinetics*, W. H. Freeman, 1967.

General textbooks

Partington, J. R., *Advanced Treatise on Physical Chemistry*, Longmans, 1949.
Taylor, H. S., and Glasstone, S., *Treatise on Physical Chemistry*, 3rd Edn., MacMillan, Vol. 1, 1942; Vol. 2, 1951.

Spectra

Herzberg, G., *Infra-red and Raman Spectra of Polyatomic Molecules*, D. van Nostrand Co., 1945.
Herzberg, G., *Spectra of Diatomic Molecules*, D. van Nostrand Co., 1950.

Kinetics

Glasstone, S., Laidler, K., and Eyring, H., *Theory of Rate Processes*, McGraw Hill Book Co., 1941.

Benson, S. W., *Foundations of Chemical Kinetics*, McGraw Hill Book Co., 1960.

Holbrooke, K., and Robinson, P. J., *Unimolecular Reactions*, Wiley, 1971.

Data sources

Benson, S. W., *Thermochemical Kinetics*, Wiley, 1968.

Janz, G. J., *Estimation of Thermodynamic Properties of Organic Compounds*, Academic Press, 1958.

Cox, J. P., and Pilcher, G., *Thermochemistry of Organic and Organometallic Compounds*, Academic Press, 1970.

Timmermanns, J., *Physicochemical Constants of Pure Organic Compounds*, Elsevier Co., Amsterdam, 1950.

Rossini, F. D., Wagner, D. D., Evans, W. H., Levine, S., and Jaffe, I., *Selected Values of Chemical Thermodynamic Properties*, National Bureau of Standards Circular 500, U.S. Government Printing Office, Washington, D.C., 1952.

Rossini, F. D., Pitzer, K. S., Arnett, R. L., Braun, R. M., and Pimentel, G. C., *Selected Values of Physical and Thermodynamic Properties of Hydrocarbons and Related Compounds*, American Petroleum Institute Research Project, Carnegie Press, Pittsburgh.

J.A.N.A.F. Tables of Thermochemical Data, Dow Chemical Company, Midland, Michigan.

Sutton, L. E. (Ed.), *Interatomic Distances*, Chemical Society of London, Special Publication, 1958; supplement 1965.

Trotman-Dickenson, A. F., and Milne, G. S., *Tables of Bimolecular Gas Reactions*, National Bureau of Standards Circular, U.S. Government Printing Office, Washington, D.C., 1967.

Benson, S. W., *Tables of Unimolecular Gas Reactions*, National Bureau of Standards Circular, U.S. Government Printing Office, Washington, D.C., 1970.

Kondratiev, V. N., *Rate Constants of Bimolecular Gas Reactions*, Academy of Sciences, U.S.S.R., Moscow, 1970.

SOLUTIONS TO PROBLEMS

1.2 (a) $4^3 = 64$ states; (b) cannonical ensemble; (c) $\epsilon + \epsilon + 4\epsilon$, 3 states; $2\epsilon + 2\epsilon + 2\epsilon$, 1 state; $\epsilon + 2\epsilon + 3\epsilon$, 6 states; (d) microcanonical ensemble.

2.1 $1 \cdot 18 \times 10^{-10}$ m.

2.2 There are 6 π-electrons. Two can have $j = 0$, and four $j = 1$, two electrons of opposite spins occupying any rotational quantum state. The lowest observable transition will then be from $j = 1$ to $j = 2$. While the transition $j = 0$ to $j = 1$ has a wavelength of 520 nm, that from $j = 1$ to $j = 2$ has a wavelength of 175 nm. In view of the crude model the correlation is as good as could be expected.

2.3 $N(\epsilon) = 1 \cdot 2 \times 10^{31}$ states J^{-1}; $n = 5 \cdot 0 \times 10^{10}$. $N(\epsilon)$ is much the larger because the unit of energy is vastly greater than kT.

2.5 Translational quantum number $= 4 \cdot 2 \times 10^8$; rotational quantum number $= 10$. The relative magnitudes reflect the difference in "box size".

4.1 All states of energy 6ϵ are equally probable, but the energy distribution $\epsilon + 2\epsilon + 3\epsilon$ is twice as probable as $\epsilon + \epsilon + 4\epsilon$, and six times as probable as $2\epsilon + 2\epsilon + 2\epsilon$.

4.2 The absolute size of the energy fluctuation decreases as the size of the assembly decreases, but the relative fluctuation $\delta E/E$ increases and for a single molecule becomes of the order of unity.

4.5 $C_V = 2kT(\partial \ln Q/\partial T)_V + kT^2 (\partial^2 \ln Q/\partial T^2)_V.$

4.7 $\sigma_E/\bar{E} = (2/3N)^{\frac{1}{2}}$. For $N = 10^{20}$, $\sigma_E/\bar{E} \approx 10^{-10}$ which is negligible. Volume of Ar at STP for a fluctuation of $10^{-6} \approx 2 \cdot 5 \times 10^{-3}$ mm^3.

5.3 ^{3}He, D, HD, ^{13}C^{16}O, ^{12}CH$_3$D, NO are fermions; ^{4}He, H, ^{19}F, ^{12}C, ^{16}O, ^{12}CH$_4$, H$_2$O are bosons.

5.6 The values for $\log_{10} W$ are respectively 59, 116, 600 and 1577.

5.7 Number of electrons per $m^3 = 2 \cdot 6 \times 10^{28}$. The number of occupied translational quantum states is half this because two electrons of opposite spin can occupy each state. The radius of the sphere of quantum number space is $2 \cdot 9 \times 10^9$, and the energy of the highest occupied quantum state $8 \cdot 5 \times 10^5 \, \text{J mol}^{-1}$, which is about $30kT$.

6.1 $q_{\text{trans}}(117 \, \text{K}) = 1 \cdot 17 \times 10^{28}, q_{\text{trans}}(1000 \, \text{K}) = 3 \cdot 13 \times 10^{29}$.

6.2 $q_{\text{trans}} = 7 \cdot 7 \times 10^{19}$.

6.3 Ethane, 6; 2 methyl propane, 3; benzene, 12; benzoquinone, 4; toluene, 1; nickel carbonyl (tetrahedral), 12; sulphur hexafluoride (octahedral) 24; dimethyl cadmium, 18.

6.4 $q_{\text{rot}}(\text{CN}) = 107 \cdot 0; q_{\text{rot}}(\text{OH}) = 11 \cdot 05, q_{\text{rot}}(\text{ClO}) = 302 \cdot 9$.

6.5 For SF_6: $ABC = 6 \cdot 83 \times 10^6 \, \text{Å}^6 \, \text{amu}^3$; $\sigma = 24$; $q_{\text{rot}} = 8 \cdot 40 \times 10^3$. For BF_3: $ABC = 2 \cdot 184 \times 10^5 \, \text{Å}^6 \, \text{amu}^3$; $\sigma = 6$; $q_{\text{rot}} = 6 \cdot 00 \times 10^3$.

6.8 There are two symmetrical stretching modes, one unsymmetrical stretching mode and two bending modes each of which is doubly degenerate.

6.9 There are three non-degenerate bending modes, two symmetrical stretching modes and one unsymmetrical stretching mode.

6.10 One of the symmetrical stretching modes becomes the reaction mode.

6.11 The vibrational partition functions for the four frequencies are $1 \cdot 1228$, $1 \cdot 5425$, $1 \cdot 0248$ and $1 \cdot 2850$. The total vibrational partition function allowing for degeneracies is $6 \cdot 1006$. $A = B = C = 294 \cdot 8 \, \text{Å}^2$ amu; $\sigma = 12$, giving $q_{\text{rot}} = 3 \cdot 25 \times 10^4$.

6.13 The difference arises from the degeneracy of the energy levels of the two dimensional rotator.

6.14 The degeneracies of the energy levels are $(n + 1)$. The result is similar to that of Problem 6.13.

6.15 $p_j = (2j + 1) \exp [-j(j + 1) \, \theta_r/T] (T/\theta_r)$. Setting $dp_j/dj = 0$ gives $(2j + 1)^2 \, (\theta_r/T) = 2$, and hence with $\theta_r/T = 0 \cdot 05$ $j_{\text{max probability}} = 2 \cdot 66$. The values for p_j for energy levels with $j = 0$ to $j = 10$ are respectively $0 \cdot 0500$, $0 \cdot 1357$, $0 \cdot 1852$, $0 \cdot 1920$, $0 \cdot 1655$, $0 \cdot 1227$, $0 \cdot 0796$, $0 \cdot 0456$, $0 \cdot 0232$, $0 \cdot 0105$, $0 \cdot 0043$.

6.16 Probabilities for occupation of rotational levels $j = 0$ to $j = 2$ are $0 \cdot 0508$, $0 \cdot 1377$, $0 \cdot 1873$; and of the vibrational levels $1 \cdot 000$, 6×10^{-7}, 4×10^{-15}. 10% of HCl molecules are in the first vibrational level at 1960 K.

6.17 The energies of the transitions are $\epsilon_{(j \, to \, j+1)} = h\nu + h^2(j + 1)/(4\pi^2 I)$ and $\epsilon_{(j \, to \, j-1)} = h\nu - h^2 j/(4\pi^2 I)$. The maximum line intensity should occur for j about 3 as shown by Problems 6.15 and 6.16.

8.2 $S_{300\,K} = 1\cdot399$ J K^{-1} mol^{-1}, $S_{1000\,K} = 18\cdot49$ J K^{-1} mol^{-1}. These are lower than the experimental values because of the inadequacy of the Einstein theory.

8.3 The high temperature limit is around 50 J K^{-1} mol^{-1} because one mole of KCl contains two moles of ions. The data give a reasonable fit with $\theta_{Einstein} = 150$ K.

9.2 The entropies in J K^{-1} mol^{-1} are Ar, 154·86; Kr, 164·09; Xe, 169·69.

9.3 $\mu_{trans} = -NkT\left(\frac{3}{2} \ln MW + \frac{5}{2} \ln T - \ln P - 3\cdot6650\right)$.

9.4 I$_2$: $S_{trans} = 188\cdot53$, $S_{rot} = 78\cdot52$, $S_{vib} = 12\cdot45$, $S_{total} = 279\cdot50$ J K^{-1} mol^{-1}.
HI: $S_{trans} = 179\cdot99$, $S_{rot} = 41\cdot371$, $S_{vib} = 0\cdot083$, $S_{total} = 221\cdot44$ J K^{-1} mol^{-1}.

9.5 BF$_3$: $S_{trans} = 161\cdot45$, $S_{rot} = 84\cdot81$, $S_{vib} = 8\cdot110$, $S_{total} = 254\cdot37$ J K^{-1} mol^{-1}.

9.6 I$_2$: $C_{vib} = 8\cdot055$, $C_V = 28\cdot84$ J K^{-1} mol^{-1}.
HI: $C_{vib} = 0\cdot478$, $C_V = 21\cdot263$ J K^{-1} mol^{-1}.

9.7 NH$_3$: $C_{vib} = 22\cdot465$, $C_V = 47\cdot407$ J K^{-1} mol^{-1}.

9.8 $q_{el} = 1 + \exp[-\theta_{el}/T]$ where $\theta_{el} = 121 \times 1\cdot4388$.
$E_{el} = 520$ J mol^{-1}; $C_{V(el)} = 0\cdot645$ J K^{-1} mol^{-1}; $S_{el} = 5\cdot40$ J K^{-1} mol^{-1}.

9.9 The condition for the maximum is $(2T_{max} + \theta) = (\theta - 2T_{max}) \exp[\theta/T_{max}]$. Numerical solution gives $\theta/T_{max} = 2\cdot40$, whence $T_{max} = 73$ K.

9.11 $I = 1\cdot605$ Å^2 amu, $\sigma = 3$. $C/R = 0\cdot98$ in good agreement with Figure 9.8.

10.1 $\bar{s} = 4\cdot45 \times 10^2$ m s^{-1}, speed of sound $= 3\cdot3 \times 10^2$ m s^{-1}.

10.2 The mean speed in the forward direction for 1 dimension is obtained by integrating $p(v_x) \, dv_x$ of equation (10.1.2) from zero to infinity. The result is $\hat{v}_x = (2kT/\pi m)^{\frac{1}{2}}$. The mean speed of Cs atoms is then 200 m s^{-1} at 1000 K.

10.3 The fraction is $2 \times 0.1 \times \exp\left[-1/\pi\right]/\pi = 0.046$. In both this and the previous example it is necessary to remember that only half the atoms are considered, those moving forwards.

10.4 The escape velocities from the earth and moon are respectively 1.12×10^4 and 2.38×10^3 m s^{-1}. The fractions of molecules which can escape are obtained as half of f_{1-D} given by equation (10.3.4). The other half are, of course, moving in the wrong direction. The fractions which can escape are

	Earth		Moon	
	H_2	N_2	H_2	N_2
Fractions	10^{-23}	10^{-305}	2×10^{-2}	10^{-15}
Rate of escape	very slow	negligible	very fast	moderately fast

Over geological aeons all H_2 may be expected to escape from the earth and all gases from the moon, hence the absence of lunar atmosphere.

10.6 $p_h/p_{\text{surface}} = \exp[-mgh/kT]$; p_h = pressure at height h above surface, m = molecular mass, g = acceleration due to gravity.

10.7 $\mu F/kT = 4 \times 10^{-5}$ which is very much less than unity.

11.1 The usual formula for K_c is in error by the factor

$$\exp\left[\sum_{\text{products}}(1/2N) - \sum_{\text{reactants}}(1/2N)\right]$$

For N large this is negligibly different from unity.

11.2 $K_{c(\text{trans})} = 0.3688$, $K_{c(\text{rot})} = 11.49$, $K_{c(\text{vib})} = 1.0119$, $K_{c(\text{el})} = 0.4383$ $\exp\left[-\Delta\epsilon_0^\circ/kT\right] = 0.5532$. Whence $K_c = 1.04$.

11.3 Following the method used for the fluorine dissociation (worked example 11.3) the heat of dissociation at 0 K is found to be 149.21 kJ mol^{-1} at 1275, and 148.99 kJ mol^{-1} at 1173, agreeing excellently with the accepted value obtained spectroscopically of 148.97 kJ mol^{-1}.

11.4 $I_{CN} = 7.78$ Å^2 amu, $\sigma = 1$; $q_{\text{vib(CN)}} = 1.439$.
$I_{C_2N_2} = 107.02$ Å^2 amu, $\sigma = 2$; $q_{\text{vib}} = 5213$.
$\exp\left[-\Delta\epsilon_0^\circ/kT\right] = 1.19 \times 10^{-10}$.
$K_p = 7.7 \times 10^{-3}$ atm in reasonable agreement with literature data.

12.4 The exponents of (q_v/q_r) are respectively 2, 5, 3 ,4 and 6.

12.5 The complex having two fewer rotations than the reactant
$$A = (kT/h)\,(q_v/q_r)^2 \approx 10^{11\cdot5}\ s^{-1}$$

12.6 $E_{exp} = E_0 - (3/2)\,RT.$

12.7 Excluding symmetry numbers $A = 3\cdot7 \times 10^7\ mol^{-1}\ m^3\ s^{-1}$. Since however the H_2 molecule can be attacked at either end the proper A-factor is double this, that is $7\cdot4 \times 10^7$, in good agreement with the experimental value.

INDEX

257